T0135658

Kernel Methods in Chemo- and Bioinformatics

Dissertation

der Fakultät für Informations- und Kognitionswissenschaften

der Eberhard-Karls-Universität Tübingen

zur Erlangung des Grades eines

Doktors der Naturwissenschaften

(Dr. rer. nat.)

vorgelegt von

Dipl.-Inf. Holger Fröhlich

aus Berlin

Tübingen
2006

Bibliografische Information der Deutschen Nationalbibliothek

Die Deutsche Nationalbibliothek verzeichnet diese Publikation in der Deutschen Nationalbibliografie; detaillierte bibliografische Daten sind im Internet über http://dnb.d-nb.de abrufbar.

ISBN 3-8325-1439-2
ISBN13 978-3-8325-1439-6

Logos Verlag Berlin
Comeniushof, Gubener Str. 47,
10243 Berlin
Tel.: +49 030 42 85 10 90
Fax: +49 030 42 85 10 92
INTERNET: http://www.logos-verlag.de

Zusammenfassung

Die vorliegende Dissertation beschäftigt sich mit Problem der Chemie- und Bioinformatik, für die Lösungsmöglichkeiten mit Hilfe spezieller maschineller Lernverfahren, sogenannter Kernelmethoden, untersucht werden. Kernelmethoden haben in den letzten Jahren auf Grund ihrer guten theoretischen Fundierung und des erfolgreichen Einsatzes in verschiedensten Bereichen ein steigendes Interesse erfahren. Gemeinsam ist allen Kernelmethoden die Verwendung einer Kernfunktion, die als ein spezielles Ähnlichkeitsmaß aufgefaßt werden kann.

Am Anfang dieser Arbeit wird ein Überblick über Grundlagen und Prinzipien kernelbasierter Lernmaschinen vermittelt. Danach wird ein neuer Algorithmus zur Modellselektion für Support Vector Machines (SVMs) in der Klassifikations- und Regressionsanalyse vorgeschlagen, der auf Ideen aus der globalen Optimierungstheorie basiert. Der Algorithmus macht keine Annahmen über spezielle Eigenschaften der Kernfunktion, wie z.B. Differenzierbarkeit, und ist sehr effizient. Experimentelle Vergleiche zu existierenden Verfahren zur Modellselektion zeigen die positiven Eigenschaften unseres Algorithmus.

Nach dem Problem der Modellselektion wendet sich die vorliegende Arbeit Anwendungen von Kernelmethoden im Bereich der Chemie- und Bioinformatik zu: Für das Problem der ADME *in silico* Vorhersage in der modernen Arzneimittelentwicklung werden Molekülrepräsentationen untersucht, die auf Deskriptoren oder auf Graphen basieren. Es wird ein Algorithmus zur Deskriptorselektion vorgeschlagen, der die statistische Stabilität eines existierenden Verfahrens verbessern soll. Ferner wird eine neue Klasse von speziellen Kernfunktionen entwickelt, die den Vergleich eines Molekülpaares auf Graphebene erlaubt. Es werden verschiedene Kombinationen von Molekülrepräsentationen basierend auf Graphen und auf Deskriptoren untersucht, die einerseits die Einbeziehung von Expertenwissen und andererseits die Integration verschiedener Ähnlichkeitsmaße in ein SVM-Modell erlauben. Ferner wird eine reduzierte Graphdarstellung von Molekülstrukturen vorgeschlagen, in der bestimme Strukturelemente in einem Knoten des Graphen zusammengefaßt werden. Experimente zeigen, daß mit unserer Methode Verbesserungen der Vorhersagefähigkeit verglichen zu derzeit üblichen Modellierungsansätzen erzielt werden können. Gleichzeitig ist unser Verfahren relativ günstig in Bezug auf den Berechnungsaufwand, sowie einheitlich und sehr flexibel einsetzbar.

Eine andere Frage, die innerhalb dieser Dissertation untersucht wird, ist, welche Eigenschaften des Membranpotenials (MP) bei kortikalen Neuronen *in vivo* die Auslösung eines Aktionspotentials (AP) bewirken. SVMs werden trainiert, um das Auftreten eines AP basierend auf mehreren Merkmale, die aus dem Verlauf des MP extrahiert werden, vorherzusagen. Ein spezieller Algorithmus zur Merkmalsselektion wird angewandt, um die wichtigsten Attribute simultan aus mehreren *in vivo* Ableitungen zu bestimmen. Die Schlußfolgerung aus unseren Experimenten ist, daß das Auftreten eines AP nicht nur von der Potentialhöhe des MP kurz vor dem Anfang des AP abhängt, sondern auch von der Änderungsrate des MP, dem Anstieg des Membranpotentials mehrere Millisekunden vor Auslösung des AP, sowie dem durchschnittlichen MP über einen längeren Zeitraum. Unsere Erkenntnisse erweitern systematisch Untersuchungen aus anderen Forschungsarbeiten und werden z.T. auch von diesen bestätigt.

Eine letzte Anwendung von Kernelmethoden in dieser Arbeit beschäftigt sich mit dem Problem, wie man Gene hinsichtlich ihrer Funktion basierend auf ihrer Gene Ontology (GO)-Annotation clustern kann. Zu diesem Zweck werden spezielle Kernelfunktionen entwickelt, welche die Ähnlichkeit zwischen Genenprodukten unter Berücksichtigung der GO-Graphstruktur bestimmen. Unter Einsatz verschiedener Clusterverfahren, wie K-Means, Spektralclustering und Average Linkage, können mit unserer Methode sinnvolle Cluster erkannt werden. Der Einsatz auf andere Ontologien oder Taxonomien ist mit dem Verfahren im Prinzip möglich.

Abstract

This thesis is devoted to the finding of possible solutions for some machine learning related problems in modern chemo- and bioinformatics by means of so-called kernel methods. They are a special family of learning algorithms that have attracted a growing interest during the last years due to their good theoretical foundation and many successful practical applications in various disciplines. At the core of all kernel methods is the usage of a kernel function, which can be thought of as a special similarity measure between arbitrary objects.

At the beginning of this thesis fundamentals and principles of kernel machines are reviewed. Afterwards a novel algorithm for model selection for Support Vector Machines (SVMs) in classification and regression is proposed, which is based on ideas from global optimization theory. It does not make any assumptions about special properties of the kernel function, like differentiability, and is highly efficient. Experimental comparisons to existing algorithms yield good results.

After this we turn our point of interest to applications of kernel methods in chemo- and bioinformatics: For the ADME *in silico* prediction problem in modern drug discovery descriptor and graph-based representations of molecules are investigated. A descriptor selection algorithm is proposed, which can improve the statistical stability of an existing method. Furthermore, a novel class of specialized kernel functions is introduced that allows the comparison of a pair of molecules on a graph-based level. Various combinations of graph and descriptor-based representations are investigated, which on one hand allow the incorporation of expert domain knowledge and on the other hand the integration of different notions of molecular similarity in one SVM model. Furthermore, a reduced graph representation for molecular structures is proposed, in which certain structural elements are condensed in one node of the graph. Our experiments indicate that with our method improvements of the prediction performance compared to state-of-the-art modeling approaches can be achieved. At the same time our method is computationally rather cheap, unified and highly flexible.

Another question, that is examined in the content of this thesis, is, which features of the membrane potential (MP) determine the generation of action potentials (APs) in cortical neurons *in vivo*. SVMs are trained to predict the occurrence of an AP before its onset based on several extracted features of the MP. A specialized feature selection algorithm is then used to select the most important features simultaneously in several *in vivo* recordings. In conclusion we find that the occurrence of an AP not only depends on the value of the MP shortly before AP onset, but also on the MP rate of change, the increase of the membrane potential several ms before AP onset, and the long range mean MP. Our findings systematically extend investigations by other researchers and are partially also confirmed by their results.

As a last application of kernel methods in this thesis, we deal with the problem of clustering genes with regard to their function based on their Gene Ontology (GO) annotation. For this purpose specialized kernel functions are developed, which measure the similarity between gene products with respect to the structure of the GO graph. Using several clustering algorithms, like kernel k-means, spectral clustering and average linkage, we can detect meaningful clusters with our method. Applications to other ontologies or taxonomies in principle are possible.

Acknowledgments

Ich möchte mich bei dem gesamten Lehrstuhl Prof. Zell bedanken für die Unterstützung und das sehr angenehme Arbeitsklima, das ich im Rahmen meiner dortigen Tätigkeit als wissenschaftlicher Mitarbeiter in den vergangenen Jahren erfahren habe. An erster Stelle möchte ich dabei natürlich meinen Chef und Betreuer Prof. Dr. Andreas Zell nennen. Ihm ist es zu verdanken, daß sein Lehrstuhl personell, von Seiten der Hardware und finanziell so gut ausgestattet ist, wodurch mir einige interessante Konferenzreisen ermöglicht wurden. Zu erwähnen wären hier besonders die ICTAI 2003 in Sacramento, die IJCNN 2004 in Budapest, die IJCNN 2005 in Montreal und die ICML 2005 in Bonn.

Mein besonderer Dank gilt ferner meinen Kollegen Jörg Wegner und Nora Speer, sowie meinem Schulfreund Björn Naundorf vom MPI für Dynamische Systeme in Göttingen, mit denen ich einige der Publikationen zusammen geschrieben habe, die entscheidend zum Entstehen dieser Dissertation beigetragen haben. Insbesondere die wissenschaftlichen Diskussionen mit Jörg über QSAR-Modelle und ADME-Vorhersage waren überaus fruchtbar.

Dankend erwähnen möchte ich auch meinen Bekannten Jean-Yves Audibert, CERTIS, Frankreich, der mir für die Beweisführung bei den "Optimal Assignment Kernels" ein paar nützliche Tips gegeben hat.

Für die Betreuung der Rechner möchte ich mich auch noch einmal bei unserem Systemadministrator Klaus Beyreuther bedanken, der stets ein offenes Ohr für alle hat und schnell und unbürokratisch hilft, wo er kann. Dank gebührt außerdem auch unserer Sekretärin Claudia Walter, ohne die der gesamte Lehrstuhlbetrieb wohl nicht funktionieren würde.

Ferner möchte ich auch noch einmal meinen Dank an Prof. Dr. Bernhard Schölkopf richten, ohne dessen Betreuung meiner Diplomarbeit im Jahr 2002 ich wohl niemals in die Geheimnisse der Kernel Maschinen eingedrungen wäre.

Meiner Freundin Andrea Brammen möchte ich meinen herzlichen Dank aussprechen für die Liebe, die sie mir all die Jahre gewährt hat. Zu guter letzt möchte ich noch meine Eltern, Großeltern und Geschwister nennen, die mir Zeit meines Lebens immer eine verläßliche Stütze waren. Dabei möchte ich besonders meinen toten Bruder Hagen erwähnen, der mir so viel gegeben hat. Ihm möchte ich daher diese Arbeit widmen.

Contents

Chapter 1

Introduction

1.1 Motivation

1.1.1 Chemo- and Bioinformatics: New Challenges for Machine Learning

Unraveling regularities in measured data is a core problem emerging in a broad range of disciplines covering economics, engineering as well as natural science. While human beings are highly specialized to deal with this task with high precision, they are at the same time limited to the processing of small amounts of data in low dimensions. Hence, computer aided pattern discovery has achieved a growing impact during the last decades. With the invention of DNA microarray technology in biology and the growing interest in virtual screening methods in modern drug design new and highly challenging tasks with massive amounts of data to be analyzed have emerged during the last years. The connection of natural and computer science makes this field a unique and fascinating area of research. New solutions have to be invented to deal with the arising problems in an appropriate way. The challenges not only lie in noisy or incomplete data measurements, but especially in difficult data structures, as e.g. DNA sequences, molecule structures or ontology graphs, which classical machine learning methods cannot directly deal with. The usual way of transforming the input structure into a "flat" vector representation, which is suitable for training a classical learning algorithm, has the disadvantage that information, which can be of great value for the problem at hand, may be difficult to encode or even be missed. Hence, the problem for machine learning in natural science applications is two-fold: First, robust, fast and precise prediction methods are needed to account for noisy, incomplete, massive or very limited amounts of data. Second, learning paradigms are preferable, which allow to encode the special structure of the data in an easy way and hence enable to integrate valuable human expert knowledge.

The range of natural science topics addressed in this thesis is relatively broad and touches different areas from natural science, such as chemistry, genome biology and neurobiology. We present a brief overview over each of them:

ADME *in Silico* Prediction The problem of ADME (*A*bsorption, *D*istribution, *M*etabolism, *E*xcretion) *in silico* prediction is of high importance in modern drug discovery (Kubinyi, 2002, 2003, 2004). The goal is to predict absorption, distribution, metabolism and excretion properties of a molecule via statistical methods purely from its structure. This way molecules, which can potentially serve as a drug later on, can be filtered out from the large rest of others, which are of no interest in an early stage of the drug discovery process. A crucial problem in this context is to find an abstract molecule description, which makes structures more similar that are akin with regard to the physico-chemical property we want to predict. This task is far from being trivial, since molecular similarity is highly context dependent (Bender and Glen, 2004). Structures, which are highly similar with regard to one property, can be dissimilar with regard to another one at the same time.

Classically *Q*uantitative *S*tructure *P*roperty *R*elationship (QSPR) models are based on a high dimensional vector representation of the molecular structures, where each or several entries of the vector describes certain physico-chemical, steric, topological or geometrical properties of the molecule (Todeschini and Consonni, 2000). It, is however, in general a-priori not totally clear, which of these descriptors are actually needed to obtain an optimal prediction performance of the learning algorithm. This is due to the fact that most machine learning algorithms are sensitive to features, which are irrelevant for the prediction task at hand (John et al., 1994; Blum and Langley, 1997; Kohavi and John, 1997; Guyon and Elisseeff, 2003). Therefore algorithms for automatic selection of relevant descriptors or features in an abstract vector representation of a molecule play a crucial role (e.g. Byvatov and Schneider, 2004). A broad range of general methods exist, but there is still a need for approaches, which take into account the special structure of the problem at hand.

Complementary to a vector representation, it is possible to view molecular structures as labeled graphs. *Feature trees* (Rarey and Dixon, 1998) are probably one of the most prominent examples for such an approach. Since any learning algorithm relies at its core on some notion of similarity or dissimilarity of the data objects, which are to be analyzed, the arising problem within a graph-based approach is, however, how the similarity or dissimilarity of molecular graphs can be defined in a computationally efficient way. Here machine learning based approaches, which allow the explicit definition of a similarity concept and do not make any assumptions on the structure of the data have a natural advantage.

Which Features Trigger Action Potentials in Cortical Neurons *in Vivo*? Up to now it is far from being clear, which features of the membrane potential (MP) of cortical neurons *in vivo* determine the occurrence of an action potential (AP). As already shown in the work by Hodgkin and Huxley (1952), the emission of an AP is a non-linear high-dimensional process involving the detailed dynamics of voltage-dependent channels in the membrane of a neuron. However, emergent mechanisms resulting from their dynamical interplay were investigated only recently (Koch et al., 1995; y Arcas et al., 2003). In these studies it was predicted that the voltage at which an AP initiates partially depends on the velocity with which the MP depolarizes. Neurons *in vivo* are subject to an immense synaptic bombardment, leading to large fluctuations of their MP. This opens the question, whether nonetheless there can be detected patterns in the sub-threshold fluctuations of the MP, which can predict the occurrence of an AP in such a natural environment. Machine learning based approaches capable of dealing with the noise in the measurements are therefore an option to address this problem.

Kernel-Based Functional Grouping of Genes After DNA microarray experiments researchers in biology usually come up with a list of differentially expressed genes, which they like to examine further. In this context they often categorize genes with regard to known biological functions, which can be obtained from the Gene Ontology (GO — The Gene Ontology Consortium, 2004). Existing methods often look for GO terms, which are statistically overrepresented in the given list of candidate genes (e.g. Beißbarth and Speed, 2004). A complementary idea is to group genes together having a similar functional profile. This is, however, obviously only possible by hand for a very small number of genes. Hence, an automatic clustering method seems to be useful, which should take into account the GO structure as a directed acyclic graph (DAG). Thereby, especially learning methods, which allow the definition of similarity concepts that incorporate this structural information, seem to be well suited.

1.1.2 Machine Learning: State of the Art

Overview Beginning from Rosenblatt's pioneering work (Rosenblatt, 1958) the field of machine learning has run through a long process of growing maturity and has today reached a stage, where in many real life applications computers can achieve a performance close to or even better than a human being (e.g. Schölkopf and Smola, 2002; Schölkopf et al., 2004). This can be mainly contributed to a growing theoretical understanding of the principles of machine learning, which resulted in better and better learning algorithms (Vapnik, 1995, 1998; Hastie et al., 2001). Over time approaches from different disciplines have come together and have resulted in new ideas. Therefore, the whole field today has reached a high diversity with a rich toolbox of available methods. During the last years, especially the following approaches have become a mainstream:

- Bayesian methods, e.g. Gaussian Processes, Bayesian networks (Heckerman, 1997; MacKay, 1997; Willams, 1997; Gelman et al., 2004).

- kernel methods, e.g. SVMs, Kernel PCA (Cristianini and Shawe-Taylor, 2000; Schölkopf and Smola, 2002; Shawe-Taylor and Cristianini, 2004).
- artificial intelligence methods, e.g. inductive logic programming, decision trees (Breiman et al., 1984; Russel and Norvig, 1995).
- neural networks, e.g. multi-layer perceptrons, self-organizing maps (Bishop, 1995; Kohonen, 1995).

Besides, also meta-learning techniques, like bagging and boosting (Freund and Shapire, 1996; Breiman, 2001), have attracted a growing interest.

Why Kernel Methods? In this thesis our focus is on kernel methods for the following reasons: Kernel methods, like the well known Support Vector Machine (SVM — Boser et al., 1992; Cortes and Vapnik, 1995), are theoretically well founded by the underlying principles of statistical learning and regularization theory (Vapnik, 1995, 1998; Tikhonov and Arsenin, 1977) and are hence relatively robust against overfitting and noisy data measurements. Kernel-based algorithms rely on the representation of pattern similarities by means of a so-called kernel function, which defines a dot product between a pair of objects in a possibly infinite dimensional Hilbert space (Vapnik, 1995, 1998; Schölkopf and Smola, 2002; Shawe-Taylor and Cristianini, 2004). A key observation is that most classical, well established methods from statistics, like Principal Component Analysis (PCA), Fisher's discriminant analysis or k-means clustering, can be reformulated in terms of kernels and thus enable nonlinear pattern analysis (Schölkopf and Smola, 2002). Additionally, a kernel function allows to deal with arbitrarily structured data, because a kernel function can be viewed as a special similarity measure, which implicitly embeds the original data into some Hilbert space. This way, using a SVM, it is possible to directly classify strings, trees or graphs without the need to construct a vector representation, which might miss some useful information provided by the original structure (e.g. Haussler, 1999; Kashima et al., 2004; Vishwanathan and Smola, 2004; Leslie et al., 2004). This property clearly distinguishes kernel-based methods from other machine learning approaches, as e.g. neural networks (Bishop, 1995) or decision trees (Breiman et al., 1984). Additionally, kernel methods, like SVMs, are relatively easy to interpret as they are basically linear methods operating in some high dimensional Hilbert space. Compared to nonlinear Bayesian methods, like Bayesian networks, kernel methods usually come up with significantly lower computation costs. Moreover, kernel-based learning algorithms usually rely on convex optimization problems and are therefore not prone to local optimal solutions (Schölkopf and Smola, 2002). The last aspect is a special advantage compared to all other machine learning approaches.

The nice theoretical properties of kernel-based algorithms together with their good success in practical applications, also compared to other methods, have attracted a high interest of researchers working in bioinformatics during the last years. This is also underlined by a recent textbook edited by Schölkopf et al. (2004), which especially emphasizes the potential of the kernel approach for learning in domains with a difficult structure, as e.g. DNA sequences, protein interaction graphs or others. Also applications in chemoinformatics have emerged lately (e.g. Byvatov et al., 2003; Byvatov and Schneider, 2004; Kashima et al., 2003), where kernel methods are appreciated due to their good prediction performance.

In conclusion of all the reasons mentioned above kernel methods seem to be a promising tool for the practical problems introduced in the last Subsection and are therefore chosen as our approach.

1.2 Own Contributions

Since we believe that kernel methods offer a good method to tackle the above described tasks from chemo- and bioinformatics, the main goal of this thesis is to point out possibilities, how this can be done in a useful and practical way. Thereby, in summary we see our contributions to the individual topics as follows:

ADME *in Silico* Prediction For the problem of ADME *in silico* prediction we investigate the usefulness of SVM-based descriptor selection by means of the RFE algorithm (Guyon et al., 2002) and show the superiority of a SVM-based descriptor selection scheme to an information theoretic filter approach and to a CART tree. We also study extensively the effect of incorporating prior knowledge on relevant descriptors into the descriptor selection process and demonstrate its potential usefulness on toy data. Furthermore, we propose an own algorithm, which aims at improving the statistical stability of RFE, and compare it systematically to the original RFE method. We show that our Rank Based Exchange Algorithm (REA) leads to a high improvement of a bound on the generalization error compared to the original RFE method.

As mentioned earlier, it is possible to represent molecular structures as labeled graphs (e.g. Rarey and Dixon, 1998). Thereby the problem arises, how the similarity of such graphs can be computed efficiently. We propose a specialized similarity measure for this purpose, which is based on the idea of a maximal weighted bipartite matching of the atoms of a pair of molecules and show how it can be converted into a valid kernel. At the same time the physico-chemical properties of each single atom are considered as well as the neighborhood in the molecular graph. We perform systematic comparisons of our approach to a descriptor-based model with and without automatic descriptor selection using RFE as well as to the recently proposed marginalized graph kernel (Kashima et al., 2003) and demonstrate that our proposed approach can significantly outperform each of those.

Later on our kernel is extended to deal with reduced graph representations, in which certain structural elements, like rings, donors or acceptors, are condensed in one single node of the graph. This allows to look at molecules at different scales of "resolution" and enables to put stress on structural features, which are relevant for the problem at hand. Our reduced graph representation does not simply average atomic features. Instead, it is "zoomed" into the nodes of the reduced graph, if necessary. Therefore, less information is lost as e.g. in "feature trees" (Rarey and Dixon, 1998). Our experiments indicate that the similarity measure for the reduced graph representation has on most datasets a comparable quality than the original one using the full molecular graph.

We further extensively investigate combinations of a graph and a vector representation of molecules. This on one hand allows the incorporation of expert domain knowledge in form of relevant descriptors and on the other hand it permits the integration of different notions of molecular similarity in one model. We demonstrate that this approach can lead to a further reduction of the prediction error.

Which Features Trigger Action Potentials in Cortical Neurons *in Vivo*? We examine, which patterns in the sub-threshold fluctuations of the MP of cortical neurons *in vivo* can best predict the occurrence of an AP. For this purpose we extract a set of several potentially interesting features of the MP and train a SVM to predict the occurrence of an AP before its onset. A specialized feature selection method is employed to simultaneously select the most important ones from 11 *in vivo* recordings from cat visual cortex. A systematic evaluation of the prediction performance of our model on all 11 recordings is performed. Finally, the relative influence of the most relevant features for the prediction is determined. In conclusion we find the following four basic factors to be of highest relevance: the value and the MP rate of change just before AP onset, an increase of the MP over a short period before AP onset (here: 2.5ms), and the average value of the MP over a longer time interval (here: 50ms). Using the features describing these factors, the occurrence of an AP can be predicted with very high accuracy on all our recordings. Thereby the feature with the largest impact is the MP rate of change just before AP onset. Our results imply that cortical neurons act as coincidence detectors, which are most sensitive to fast changes of the MP. Models of cortical neurons, which assume a voltage threshold for AP initiation thus might not reflect the dynamics of AP generation in the right way.

Kernel-Based Functional Grouping of Genes We introduce a method that can automatically cluster genes with regard to their function solely based on information provided by the Gene Ontology (GO — The Gene Ontology Consortium, 2004). We achieve this by means of specialized kernel functions, which measure the functional similarity of genes based on an information theoretic similarity concept for individual GO terms. We demonstrate that by using our similarity measure in combination with a k-means or spectral clustering algorithm (Ng et al., 2002) we are able to detect biologically plausible functional groups of genes.

Model Selection for SVMs Building a SVM model in the above applications always involves the tuning of certain parameters. In the context of this thesis we deal with this important machine learning problem by means of a novel algorithm, which is based on ideas from global optimization theory. In contrast to many existing methods (e.g. Chapelle et al., 2002; Chapelle and Vapnik, 2000), it does not make any assumptions about special properties of the kernel function, like differentiability, and is highly efficient. On several public available and well known benchmark datasets we perform extensive comparisons to a simple grid search as well as to the pattern search algorithm introduced by Momma and Bennett (2002), which is as general as our method. We show that our proposed method offers a speed up of factor 10 in a classification and of factor 100 in a regression experiment compared to a grid search. At the same time it has a higher statistical robustness and stability then the pattern search algorithm.

A special challenge in the context of this thesis comes from the fact that for all problems presented here a lot of domain knowledge is required. At the same time it is this combination of knowledge from different disciplines that makes this kind of applied machine learning such a highly interesting area of research.

1.3 Organization of this Thesis

The organization of this thesis is a follows: To make this work more self contained, in the next Chapter we begin with a short overview over the basic theoretical concepts that are necessary to understand the kernel-based algorithms employed in this work. Namely we depict some elements of statistical learning and regularization theory and introduce the concept of kernel functions.

Kernel-based learning algorithms, including SVMs for classification and regression (Vapnik, 1995), Kernel Ridge Regression (e.g. Shawe-Taylor and Cristianini, 2004) and others, which are utilized in this thesis, are reviewed afterwards in Chapter 3.

Beginning from Chapter 4 we present our own work: In Chapter 4 we describe a new and very efficient algorithm to perform model selection for SVMs based on strategies from global optimization theory. This finishes the first part of this thesis, which is primarily concerned with kernel algorithms itself.

In the second part various applications are presented: Chapter 5 deals with the descriptor selection problem in QSAR/QSPR models and introduces a method that is aimed to improve the statistical stability of the solution computed by the RFE descriptor selection algorithm (Guyon et al., 2002). Additionally, the influence of incorporating expert knowledge into the descriptor selection process is investigated systematically.

In Chapter 6 we propose an alternative way of dealing with QSAR/QSPR models by means of kernel functions for chemical graphs. This method is extended later on to include a-priori knowledge on relevant descriptors as well as so-called reduced graph representations.

In Chapter 7 we describe a SVM-based application that aims at finding relevant features of the membrane potential, which trigger the generation of action potentials in cortical neurons *in vivo*.

In Chapter 8 we propose a method to cluster genes according to their function based on the Gene Ontology (The Gene Ontology Consortium, 2004). For this purpose specialized kernel functions are considered.

Finally, in Chapter 9 we draw the conclusions from this work.

Chapter 2

Theoretical Foundations

2.1 Introduction

The goal of every learning algorithm is to uncover statistical relationships in data. Thereby it is important that these relationships do not hold in collected training data only, but also in previously unseen data that come from the same source. That means the output of the learning algorithm should not be over-sensitive to the particular training dataset. This property is referred to as *statistical stability* of the algorithm (e.g. Shawe-Taylor and Cristianini, 2004). At the same time we must be aware of the fact, that in many real-life applications our observed data is corrupted with noise. By noise we mean that the values of the features for individual data items may be affected by measurement inaccuracies or even miscodings, for example through human error. We wish that small amounts of noise should not affect the output of the learning algorithm too much. This property is referred to as *robustness* of the algorithm (e.g. Shawe-Taylor and Cristianini, 2004).

In statistical decision theory one is concerned with finding principles that guarantee statistical stability and robustness of a learning algorithm. Statistical decision theory comes in two basic directions: *statistical learning theory*, also called *VC (Vapnik-Chervonenkis) theory* (Vapnik, 1979, 1995, 1998), on one hand, and *Bayesian decision theory* (e.g. Bishop, 1995; Duda et al., 2001; Hastie et al., 2001) on the other hand. In this Chapter we concentrate on VC theory only, since it is the theoretical basis of most present kernel-based machine learning algorithms, like the well known Support Vector Machine (Boser et al., 1992; Cortes and Vapnik, 1995). In the following we briefly review some fundamental ideas that are necessary do understand these algorithms. A more detailed treatment can e.g. be found in the books by Vapnik (1998) and Schölkopf and Smola (2002).

We begin with a brief depiction of the basic problem for any machine learning algorithm, namely the detection of statistically stable patterns in training data. For this purpose the relation between empirical risk minimization and the theoretical generalization error is discussed in Section 2.2. This will lead us to the structural risk minimization principle described in Section 2.3, in which not only the training error induced by some learned function is considered, but also its complexity. Practically this can be achieved by regularizing the training error with a term describing the smoothness of the function to be learned in a so-called Reproducing Kernel Hilbert Space, which is a Hilbert space of functions, that is spanned by kernel functions (Section 2.4). Kernels are then discussed in greater detail in Section 2.5. Finally, in Section 2.6 we summarize.

2.2 Empirical and Expected Risk

The basic assumption in statistical learning theory is that the training data are generated by sampling i.i.d. (independent and identically distributed) from an unknown but fixed probability distribution $\mathcal{P}$. It should be noted that this assumption is actually stronger than it might appear at the first glance. Indeed, in many real-life applications it is not fulfilled (think e.g. about the expression level of a gene measured over time in a DNA microarray experiment — e.g. Eisen et al., 1998). Nevertheless, it has been turned out that this simplification does not hurt too much in practice, but offers many theoretical and algorithmic advantages.

In this Chapter we are only dealing with the case of fully supervised learning, but in principle similar arguments as presented here can be made for unsupervised learning (e.g. Principal Component Analysis) as well (c.f. textbooks by Vapnik, 1998; Schölkopf and Smola, 2002; Shawe-Taylor and Cristianini, 2004). We thus assume our training data $\mathcal{D} \sim \mathcal{P}(x, y)$ consists of n pairs $\{(x_i, y_i)\}_{i=1}^n \subset \mathcal{X} \times \mathcal{Y}$ of objects[1] x_i coming from some domain $\mathcal{X}$ (not necessarily a vector space) and labeled with $y_i \in \mathcal{Y}$. In the particular cases of a classification task $\mathcal{Y}$ would be finite (i.e. there exists a bijection $b : \mathcal{Y} \to \{1, ..., k\}$), whereas in a typical regression task we have $\mathcal{Y} = \mathbb{R}$. Based on our data $\mathcal{D}$ we wish to construct a function or *hypothesis* $f : \mathcal{X} \to \mathcal{Y}$, such that the *expected risk* or *generalization error*

$$R[f] := \int_{\mathcal{X}\times\mathcal{Y}} \ell(y, f(x))d\mathcal{P}(x, y) = \mathbf{E}_{\mathcal{D}\sim\mathcal{P}(x,y)}[\ell(y, f(x)] \tag{2.1}$$

is minimal (c.f. Vapnik, 1995, 1998). Here $\ell : \mathcal{Y} \times \mathcal{Y} \to \mathbb{R}$ denotes some *loss function* measuring the error we make by predicting $f(x)$ instead of y. A natural loss function for classification tasks is the 0-1 loss function

$$\ell(y, \hat{y}) = \begin{cases} 0 & y = \hat{y} \\ 1 & \text{otherwise} \end{cases} \tag{2.2}$$

[1]Throughout this thesis we denote objects with bold letters only, if they are explicitly declared as vectors or matrices. The component i of a vector $\mathbf{v}$ will be identified with v_i. Likewise, we write the (i, j)th component of a matrix $\mathbf{M}$ as M_{ij}. As an exception from this we denote e.g. the (i, j)th component of the inverse of $\mathbf{M}$ as $(\mathbf{M}^{-1})_{ij}$.

while for regression often the squared loss

$$\ell(y, \hat{y}) = (y - \hat{y})^2 \tag{2.3}$$

is utilized. The problem is that the quantity (2.1) cannot be computed as we do not have full access to the distribution $\mathcal{P}(x, y)$. Hence, the only way to proceed is to approximate the integral by a finite sum. This leads to the so-called *empirical risk* or training error:

$$R_{emp}[f] := \frac{1}{n} \sum_{i=1}^{n} \ell(y_i, f(x_i)) \tag{2.4}$$

The empirical risk, like the expected risk, is a functional of the unknown function f, but unlike the first one it can be calculated from the training data only, and hence a function f^n minimizing R_{emp} can be computed. A crucial question is, however, whether f^n would be also a minimizer of the expected risk as well. Due to Hoeffding's inequality[2] (Hoeffding, 1963) one can indeed ensure that for any **fixed** f the empirical risk will converge exponentially fast against the expected risk in probability, i.e. $R_{emp}[f]$ is a *statistically consistent* estimator of $R[f]$. However, for a learning machine it would obviously not make much sense to implement a fixed function f only. Instead we should assume f to be picked from some class of functions $\mathcal{C}$ that the learning algorithm can implement. A crucial insight by Vapnik and Chervonenkis (1968) is that in this case uniform convergence over the whole class of functions $\mathcal{C}$ would be required to further ensure statistical consistency of the empirical risk minimization principle:

Theorem 2.2.1 (Vapnik and Chervonenkis, 1968). *: One-sided uniform convergence in probability*

$$\lim_{n \to \infty} P(\sup_{f \in \mathcal{C}}(R[f] - R_{emp}[f]) > \epsilon) = 0 \tag{2.5}$$

for all $\epsilon > 0$ is a necessary and sufficient condition for consistency of empirical risk minimization.

Obviously, it seems extremely difficult to find a learning algorithm that ensures uniform convergence over the whole function class that can be implemented by it in practice. A way out is given by bounding $P(\sup_{f \in \mathcal{C}}(R[f] - R_{emp}[f]) > \epsilon)$ in an appropriate way. Vapnik and Chervonenkis (1968) showed that

$$P(\sup_{f \in \mathcal{C}}(R[f] - R_{emp}[f]) \leq 4 \exp\left(\log \mathbf{E}[c(\mathcal{C})] - \frac{n\epsilon^2}{8}\right) \tag{2.6}$$

[2]**Hoeffding's inequality**: If $X_1, ..., X_m$ are independent random variables satisfying $X_i \in [a_i, b_i]$, and if we define the random variable $S_m = \sum_{i=1}^{m} X_i$, then it follows for all $\epsilon > 0$ that

$$P(|S_m - \mathbf{E}[S_m]| > \epsilon) \leq 2 \exp\left(-\frac{2\epsilon^2}{\sum_{i=1}^{m}(b_i - a_i)^2}\right)$$

.

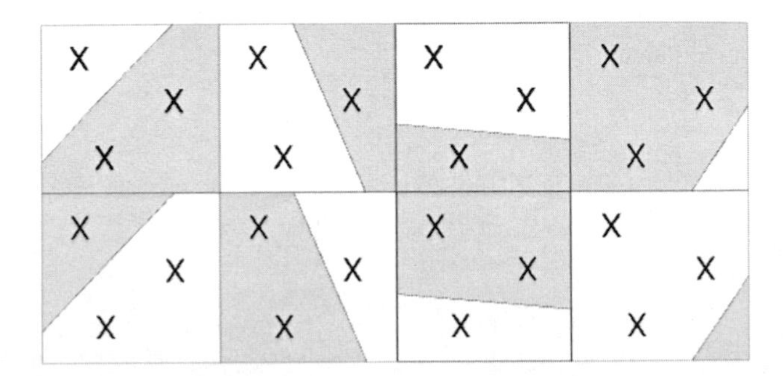

Figure 2.1: Basic idea of the VC dimension: There are $2^3 = 8$ possible ways of assigning 3 points to classes +1 and -1. In $\mathbb{R}^2$ all 8 possibilities can be realized using separating hyperplanes, i.e. the class of hyperplanes can shatter 3 points in $\mathbb{R}^2$. This would not work for 4 points, no matter how we placed them. Therefore the VC dimension of the class of separating hyperplanes in $\mathbb{R}^2$ is 3 (c.f. Schölkopf and Smola, 2002).

where $c(\mathcal{C})$ is a coefficient measuring the complexity or *capacity* of hypothesis class $\mathcal{C}$. Probably one of the most intuitive and simple capacity concepts is the so-called *VC dimension* for binary classification (c.f. Vapnik, 1979, 1995, 1998). The VC dimension is defined as the largest number l, such that there exists a set of l training points, which a given class of functions can *shatter*, and ∞ if no such l exists. A set of l points is said to be shattered by a class of functions $\mathcal{C}$, if one can realize all possible separations of the points, which are induced by any possible combinations of labeling the patterns by means of a function from $\mathcal{C}$. For example, if there are two classes +1 and -1, there are $2^3 = 8$ possible ways to label 3 training patterns. Supposed all training patterns are in $\mathbb{R}^2$ and are non-collinear each possibility can be realized using a separating hyperplane. That means the class of hyperplanes is able to shatter 3 points. This would not work for 4 points, no matter how we placed them (Fig. 2.1). Therefore the VC dimension of the class of hyperplanes in $\mathbb{R}^2$ is 3.

2.3 The Structural Risk Minimization Principle

Often the bound (2.6) is written in a slightly different form by setting the right hand side equal to some $\delta > 0$, and then solving for ϵ (e.g. Schölkopf and Smola, 2002). As a result, we get the statement that with probability of at least $1 - \delta$

$$R[f] \leq R_{emp}[f] + \sqrt{\frac{8}{n}\left(\log \mathbf{E}[c(\mathcal{C})] + \log \frac{4}{\delta}\right)} \tag{2.7}$$

From this bound the basic problem of inductive learning is revealed: While on one hand it is always possible to find a function f yielding a very low empirical risk, on the other hand for this purpose we might need a function that is a member of a very rich function class, i.e. a function class with a high capacity. This is the typical situation of *overfitting*. The learned hypothesis covers very well all special cases provided by our training examples $\mathcal{D}$, but is somehow not general enough to cover other data. Vice versa, we could choose a function of low complexity but eventually pay with a high empirical risk. We therefore *underfit* the ideal function f^* minimizing the expected risk, i.e. our learned function neither explains our training data nor unseen test examples sufficiently. The problem is to find a function that on one hand explains the given data $\mathcal{D}$ with a sufficiently low empirical risk, but on the other belongs to a function class with a capacity as low as possible. That is, informally speaking, we seek for the simplest possible hypothesis that can explain our training data.

To find such a function Vapnik and Chervonenkis (c.f. Vapnik, 1979, 1995, 1998) proposed minimizing the right hand side of (2.7) rather than concentrating on the empirical risk only. Thereby it is important to note that the capacity term $c(\mathcal{C})$ is not a property of an individual function f only, but of the function class $\mathcal{C}$. Thus, the bound cannot simply be minimized over choices of f, but has to be minimized over choices of the hypothesis class $\mathcal{C}$, which f belongs to. This leads to the *Structural Risk Minimization* (SRM) principle: The basic idea is to build a nested sequence of function classes (or *structures*) $S_1 \subset S_2 \subset ... \subset S_h \subset ...$ of increasing size (and thus, of increasing capacity) and minimize the right hand side of (2.7) over the choice of the structure. This way a function f^* is picked, which has a small training error and is an element of a structure that has a low capacity h^* (Fig. 2.2).

It is worth mentioning that the SRM principle is closely related to the so-called *bias-variance dilemma* in classical statistics (c.f. Duda et al., 2001): If in a regression example we computed a linear fit for every dataset that we ever encountered, then we could only "discover" linear functional dependencies. However, this would not be due to the nature of the data, but due to a *bias* imposed by us. If, on the other hand, we fitted a polynomial of sufficiently high degree to any given dataset, we would always be able to fit the training data perfectly, but the model would be subject to large fluctuations. The model would hence suffer from a large *variance*.

2.4 Regularization in Reproducing Kernel Hilbert Spaces

Practically it might be difficult to implement the SRM principle efficiently. Thus an approximation scheme is beneficial. The *regularization* framework (Tikhonov and Arsenin, 1977) was invented independently from VC theory, but can be understood as such an approximation. It is employed in many existing machine learning algorithms, including Support Vector Machines (Boser et al., 1992; Cortes and Vapnik, 1995; see next Chapter).

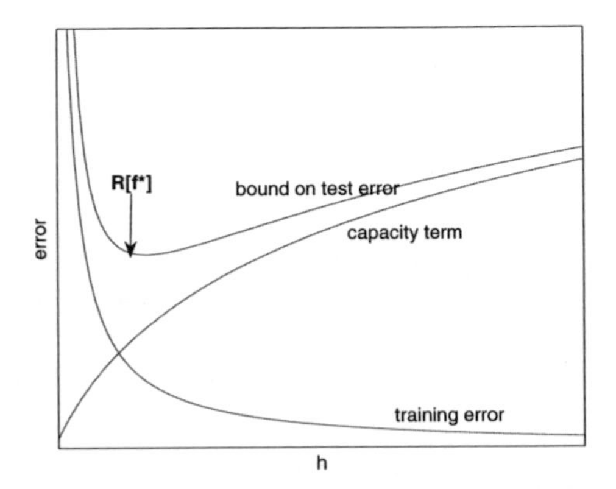

Figure 2.2: The Structural Risk Minimization (SRM) induction principle: The function class is decomposed into a nested sequence of subsets of increasing size (and thus of increasing capacity). The SRM principle picks a function f^* which has a small training error and comes from an element of the structure that has low capacity h^*, and hence minimizes a bound on the risk (c.f. Schölkopf and Smola, 2002).

The idea is to deal with the *ill-posed problem* of finding a statistically stable and robust hypothesis f by adding to the empirical risk a *regularization term* $\Omega[f]$ that involves the capacity of the function class of f. This leads to the *regularized risk*

$$R_{reg}[f] := R_{emp}[f] + \lambda\Omega[f] \tag{2.8}$$

where $\lambda > 0$ specifies the trade-off between the minimization of the complexity term and the training error. The whole functional $R_{reg}[f]$ is required to be continuous in f. Note that this is a relatively strong assumption, since it prevents us from using e.g. the 0-1 loss function (2.2) for classification problems. Instead, some continuous approximation has to be used.

Usually one chooses $\Omega[f]$ to be convex, since this ensures the minimum of (2.8) to be unique, provided that $R_{emp}[f]$ is also convex. An especially interesting case is when $\Omega[f] = \|f\|_{\mathcal{F}}^2$, the square norm of f in a *Reproducing Kernel Hilbert Space* (RKHS — Aronszajn, 1950): A RKHS is a Hilbert space of functions $f : \mathcal{X} \to \mathbb{R}$ endowed with a dot product $\langle \cdot, \cdot \rangle_{\mathcal{F}}$ and the induced norm $\|f\|_{\mathcal{F}} := \sqrt{\langle f, f \rangle_{\mathcal{F}}}$. Furthermore, there exists a so-called *kernel* function $k : \mathcal{X} \times \mathcal{X} \to \mathbb{R}$ with the following properties:

1. k has the reproducing property $\langle f, k(x, \cdot) \rangle_{\mathcal{F}} = f(x)$ for all $x \in \mathcal{X}$, in particular $\langle k(x, \cdot), k(x', \cdot) \rangle_{\mathcal{F}} = k(x, x')$.

2. k spans $\mathcal{F}$, i.e. each $f \in \mathcal{F}$ can be written as $f(\cdot) = \sum_{i=1}^{m} \alpha_i k(x_i, \cdot), m \in \mathbb{N}, \alpha_i \in \mathbb{R}, i = 1, ..., m$.

The concept of a kernel function will be treated in the next Section in more detail, but for now it is sufficient to view kernels as functions that span a RKHS. Vice versa, each RKHS uniquely determines a kernel k (Schölkopf and Smola, 2002).

It is understood that a small function norm $\|f\|_{\mathcal{F}}$ in RKHS corresponds to a *smooth* function in domain $\mathcal{X}$ (Schölkopf and Smola, 2002), which justifies the setting $\Omega[f] = \|f\|_{\mathcal{F}}^2$. Due to the *representer theorem* (Kimeldorf and Wahba, 1971) in this case the minimizer $\hat{f}$ of $R_{reg}[f]$ can be written down in closed form:

Theorem 2.4.1 (Representer Theorem, Kimeldorf and Wahba, 1971). *Denote by* $\Omega : [0, \infty) \rightarrow \mathbb{R}$ *a strictly monotonic increasing function, by* $\mathcal{X}$ *a set, and by* $\ell : \mathbb{R} \times \mathbb{R} \rightarrow \mathbb{R} \cup \{\infty\}$ *an arbitrary loss function. Then each minimizer* $\hat{f} \in \mathcal{F}$ *of the regularized risk admits a representation of the form*

$$\hat{f}(x) = \sum_{i=1}^{n} \alpha_i k(x_i, x) \tag{2.9}$$

That means the optimal hypothesis lies in the span of the $k(x_i, \cdot)$ in the RKHS. This generic form of the solution can be found in many kernel-based machine learning algorithms. It is particularly useful, because the expansion of $\hat{f}(x)$ depends on the kernels between x and the training data only. We will now look at the concept of kernel functions in greater detail in order to highlight some core properties.

2.5 Kernel Functions

We did not make any assumptions about the nature of the domain $\mathcal{X}$ from which the objects $x_i, i = 1, ..., n$ in our training data are drawn. Indeed, later on in this thesis we will deal with objects that are not represented as vectors, but for instances as graphs. However, for a practical learning algorithm it is useful to operate in some vector space. We therefore assume that there exists an embedding $\phi : \mathcal{X} \rightarrow \mathcal{H}$ of the objects coming from domain $\mathcal{X}$ into some *feature space* $\mathcal{H}$. For practical reasons it is supposed that $\mathcal{H}$ is a Hilbert space, i.e. a complete, separable[3] vector space endowed with a dot product $\langle \cdot, \cdot \rangle$. A kernel function $k : \mathcal{X} \times \mathcal{X} \rightarrow \mathbb{R}$ is then defined as a dot product

$$k(x, x') := \langle \phi(x), \phi(x') \rangle \tag{2.10}$$

[3] A vector space $\mathcal{H}$ is called *complete*, if each Cauchy sequence of elements of $\mathcal{H}$ converges to an element of it. It is *separable*, if for any $\epsilon > 0$ there is a finite set of elements $e_1, ..., e_N \in \mathcal{H}$ such that for all $e \in \mathcal{H}$ $\min_i \|e_i - e\| < \epsilon$.

of images $\phi(x), \phi(x')$ in feature space (e.g. Schölkopf and Smola, 2002; Shawe-Taylor and Cristianini, 2004). We will show later on, how this definition of a kernel function fits together with the more abstract one given in the last Section.

As k is a dot product, it can be viewed as a natural similarity measure for x, x'. Clearly, each kernel function is symmetric, since it defines a dot product in $\mathcal{H}$. Moreover, the *kernel* or *Gram matrix* $\mathbf{K} = (k(x_i, x_j))_{i,j=1}^n \in \mathbb{R}^{n\times n}$ is positive semidefinite[4], because for any $\alpha = (\alpha_1, ..., \alpha_n)^T \in \mathbb{R}^n$:

$$\begin{aligned}
\alpha^T\mathbf{K}\alpha &= \sum_{i,j=1}^n \alpha_i\alpha_j K_{ij} \qquad (2.11)\\
&= \sum_{i,j=1}^n \alpha_i\alpha_j \langle\phi(x_i), \phi(x_j)\rangle\\
&= \left\langle \sum_{i=1}^n \alpha_i\phi(x_i), \sum_{j=1}^n \alpha_j\phi(x_j)\right\rangle\\
&= \left\| \sum_{i=1}^n \alpha_i\phi(x_i)\right\|^2\\
&\geq 0
\end{aligned}$$

Vice versa, any function k that induces a symmetric, positive semidefinite matrix on a finite data set (in the following we will shortly talk about k as a symmetric, positive definite function) is a valid kernel function, i.e. it can be written as a dot product $k(x, x') = \langle\phi(x), \phi(x')\rangle$. This can be shown by constructing a RKHS $\mathcal{F}$ induced by k (e.g. Shawe-Taylor and Cristianini, 2004): For this purpose we consider the special case where $\phi(x) = k(x, \cdot)$, i.e. x is represented by its similarity to **all** other objects in feature space. This is the so-called *reproducing kernel map* (Schölkopf and Smola, 2002). We now consider functions of the set

$$\mathcal{F} := \left\{ \sum_{i=1}^m \alpha_i k(x_i, \cdot) \middle| m \in \mathbb{N}, x_i \in \mathcal{X}, \alpha_i \in \mathbb{R}, i = 1, ..., m \right\} \qquad (2.12)$$

and define for all $f, g \in \mathcal{F}$

$$\langle f, g\rangle_{\mathcal{F}} := \sum_{i=1}^n \sum_{j=1}^m \alpha_i\beta_j k(x_i, x_j) \qquad (2.13)$$

which is a well-defined dot product, since it is symmetric and

$$\langle f, f\rangle_{\mathcal{F}} = \|f\|_{\mathcal{F}}^2 = \sum_{i=1}^n \sum_{j=1}^n \alpha_i\alpha_j k(x_i, x_j) = \alpha^T\mathbf{K}\alpha \geq 0 \qquad (2.14)$$

[4]A symmetric matrix $\mathbf{M} \in \mathbb{R}^{n\times n}$ is called *positive semidefinite*, if for all $\mathbf{v} \in \mathbb{R}^n$ $\mathbf{v}^T\mathbf{M}\mathbf{v} \geq 0$. This is exactly the case, if and only if $\mathbf{M}$ has only non-negative eigenvalues.

We then have

$$\langle \phi(x), f \rangle_{\mathcal{F}} = \langle k(x, \cdot), f \rangle_{\mathcal{F}} = \sum_{i=1}^{n} \alpha_i k(x_i, x) = f(x) \tag{2.15}$$

which is exactly the reproducing property of a RKHS defined in the last Chapter. In particular we recover

$$\langle k(x, \cdot), k(x', \cdot) \rangle_{\mathcal{F}} = \langle \phi(x), \phi(x') \rangle = k(x, x') \tag{2.16}$$

which proves that indeed any symmetric, positive semidefinite function can be viewed as a dot product in some feature space. It is further easy to show the remaining separability and completeness properties of a Hilbert space for $\mathcal{F}$. Hence, each kernel k gives rise to a RKHS (c.f. Schölkopf and Smola, 2002; Shawe-Taylor and Cristianini, 2004). As already mentioned above, vice versa each RKHS uniquely determines a kernel function k (Schölkopf and Smola, 2002). Consequently, both definitions of kernels, the one given in the last Section and the one given in this Section, are equivalent.

In a different way as presented here it can also be derived from *Mercer's theorem*[5] (Mercer, 1909) that positive semi-definiteness and symmetry are necessary and sufficient conditions for any kernel k. Hence, in the literature often the term *Mercer kernels* is used. The symmetry and positive semi-definiteness properties of any Mercer kernel have a very practical implication: They allow us to check, whether a given similarity measure k is actually a kernel function and thus corresponds to a dot product in some feature space without the need to explicitly construct the feature map ϕ. This will play a crucial role throughout thesis, where we define similarity measures for arbitrary objects and prove that they are indeed Mercer kernels.

If a given function k is a symmetric, positive semidefinite kernel, then, as demonstrated above (Eq.: 2.16), the map ϕ is defined implicitly by the dot product that k computes. This is known as the so-called *kernel trick* (Aizerman et al., 1964; Cortes and Vapnik, 1995). It has a particularly useful application, if the domain $\mathcal{X} \subset \mathbb{R}^D$. In this

[5]**Mercer's Theorem:** *Let $\mathcal{X}$ be a compact subset of $\mathbb{R}^n$. Suppose k is a continuous symmetric function such that the integral operator $T_k : L_2(\mathcal{X}) \rightarrow L_2(\mathcal{X})$, $(T_k f)(\cdot) = \int_{\mathcal{X}} k(\cdot, x) f(x) dx$ is non-negative, that is*

$$\int_{\mathcal{X} \times \mathcal{X}} k(x, z) f(x) f(z) dx dz \geq 0$$

for all $f \in L_2(\mathcal{X})$. Then we can expand $k(x, z)$ in a uniformly convergent series (on $\mathcal{X} \times \mathcal{X}$) in terms of functions ϕ_j, satisfying $\langle \phi_j, \phi_i \rangle = \delta_{ij}$

$$k(x, z) = \sum_{j=1}^{\infty} \phi_j(x) \phi_j(z)$$

Furthermore, the series $\sum_{i=1}^{\infty} \|\phi_i\|^2_{L_2(\mathcal{X})}$ is convergent.

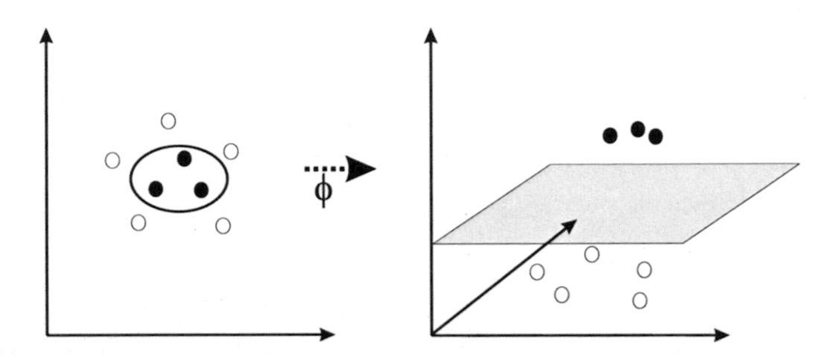

Figure 2.3: The original data $\{\mathbf{x}_i\}$ is mapped to a higher dimensional feature space, in which the images $\{\phi(\mathbf{x}_i)\}$ of the input patterns allow a linear separation of the black and the white dots. The feature map ϕ can be computed implicitly by means of a kernel function.

case we could, for instances, define $\phi(\mathbf{x})$ as all degree d monomials $x_{j_1} \cdot x_{j_2} \cdots x_{j_d}$ of components x_j of $\mathbf{x} = (x_1, ..., x_D)^T$ where $j_1, ..., j_d \in \{1, ..., d\}$. The problem is that for D-dimensional input patterns $\mathbf{x}$ there exist $\begin{pmatrix} d+D-1 \\ d \end{pmatrix}$ different monomials of degree d. For instance, if $D = 256$ and $d = 5$, then we would have a dimension of almost 10^{10}. This would make any explicit computation of dot products in this space prohibitively slow. By the usage of kernel functions, however, the same computation can be carried out without the need to explicitly calculate $\phi(\mathbf{x})$ or even knowing it and thus very efficiently. In fact, it can be shown that in our example the corresponding kernel functions are given by the dth degree *polynomial kernel*s (e.g. Schölkopf and Smola, 2002)

$$k(\mathbf{x}, \mathbf{x}') := \langle \mathbf{x}, \mathbf{x}' \rangle^d \tag{2.17}$$

which trivially also include the linear kernel $k(\mathbf{x}, \mathbf{x}') = \langle \mathbf{x}, \mathbf{x}' \rangle$ as a special case. In the latter case the feature map ϕ is the identity, while for $d > 1$ ϕ is a nonlinear map of the data lying in input space $\mathbb{R}^D$ to the higher dimensional feature space $\mathcal{H}$ (Fig. 2.3). In the extreme case $\mathcal{H}$ can even be infinite dimensional, if we consider e.g. the *RBF-kernels* of width σ

$$k(\mathbf{x}, \mathbf{x}') := \exp\left(-\frac{\|\mathbf{x} - \mathbf{x}'\|^2}{2\sigma^2}\right) \tag{2.18}$$

In this case the feature map is given by all possible monomials of input features with no restriction placed on the degree (e.g. Schölkopf and Smola, 2002).

A particular useful property of kernel functions is that they are closed under addition, multiplication, multiplication with a positive constant and exponentiation (c.f. Schölkopf and Smola, 2002; Shawe-Taylor and Cristianini, 2004). This allows the construction of new kernels from existing ones, which is especially interesting for the definition of kernels for structured objects, like strings, graphs, trees and others (e.g. Haussler, 1999; Gärtner et al., 2003; Kashima et al., 2003; Leslie et al., 2004; Vishwanathan and Smola, 2004).

2.6 Summary

We presented the basic theoretical tools that are essential to understand the algorithms employed in this thesis: We began with the observation that empirical risk minimization only allows learning a hypothesis that generalizes well, if at the same time the function class, of which the hypothesis is an element, is restricted with regard to its complexity. This lead to the SRM principle, which at the same time aims to find a function that induces a low training error and belongs to a hypothesis class with a low capacity. Practically the SRM principle can be approximated efficiently by means of the Tikhonov regularization framework. A complexity term describing the smoothness of the function to be learned is added to the empirical risk, and a parameter specifies the trade-off between low training error and low complexity. The smoothness condition can be ensured by a small norm of the function in a RKHS, which is spanned by kernel functions. Moreover, the representer theorem guarantees that each function minimizing the regularized risk functional can be written down in closed form as a linear combination of kernels evaluated at the training data only.

Equivalently to viewing kernel functions as base functions of a RKHS, kernels can be understood as dot products in some feature space and thus as natural similarity measures for arbitrary objects. It can be shown that a given symmetric similarity function is a kernel if and only if the induced kernel matrix on the training data is positive semidefinite. The dot product is then computed implicitly, which is known as the kernel trick. Important examples of kernel functions in vector spaces are polynomial and RBF kernels. A particular useful property of kernels is that they are closed under addition, multiplication, multiplication with a positive constant and exponentiation. This allows the systematic construction of new kernels from existing ones.

Chapter 3

Kernel Methods for Pattern Analysis

3.1 Introduction

In the last Chapter we introduced the theoretical foundations of kernel methods. The two key ideas are regularization and the usage of a kernel function to allow for nonlinear classification and to deal with arbitrarily structured data. The goal of this Chapter is to give an overview over those kernel-based algorithms, which are utilized in this thesis. We start with SVMs for classification in Section 3.2. We depict the basic idea of hard margin SVMs and present two variants of soft margin SVMs, namely the C-SVM and the ν-SVM. We also highlight the connections to the Tikhonov regularization framework. Furthermore, we elucidate two methods for estimating the generalization performance of a SVM efficiently solely based on the training data. Finally, we briefly review basic methods for multi-class classification with SVMs.

In Sections 3.3 and 3.4 we deal with kernel-based regression estimation: While Kernel Ridge Regression, presented in Section 3.3, is a relatively simple and straight forward method, it lacks the benefit of sparsity of the solution. This is one of the main features of Support Vector Regression, a variant of SVMs for regression, which is described in Section 3.4.

Sections 3.5 and 3.6 are concerned with unsupervised kernel methods: Kernel Canonical Correlation Analysis (KCCA — Lai and Fyfe, 2000; Akaho, 2001; Bach and Jordan, 2002), presented in Section 3.5, is a means for extracting common structure from data, which comes in two different representations or views. Section 3.6 explains some kernel-based clustering techniques. This includes dual k-means (e.g. Shawe-Taylor and Cristianini, 2004), linkage algorithms (e.g. Jain et al., 1999) in feature space and spectral clustering (Weiss, 1999; Shi and Malik, 2000; Meila and Shi, 2001a; Kannan and Vempala, 2000; Meila and Shi, 2001b; Ng et al., 2002).

3.2 Support Vector Machines for Classification

3.2.1 Hard Margin Support Vector Machines

In the last Chapter the importance of capacity control for any learning algorithm was described. Therefore, it is beneficial to come up with a class of functions, for which this can be done relatively easy, for instance the class of hyperplanes in some Hilbert space $\mathcal{H}$ with a defined dot product $\langle \cdot, \cdot \rangle$:

$$\{\mathbf{x} \in \mathcal{H} | \langle \mathbf{w}, \mathbf{x} \rangle + b = 0\}, \ \mathbf{w} \in \mathcal{H}, b \in \mathbb{R} \tag{3.1}$$

As any hyperplane divides the space into two half spaces (and thus induces a classification into a class -1 and a class +1) this corresponds to decision functions

$$f_{\mathbf{w}}(\mathbf{x}) = \operatorname{sgn}(\langle \mathbf{w}, \mathbf{x} \rangle + b) \tag{3.2}$$

Thereby one has the freedom to scale $\mathbf{w}$ by any constant factor to recover an equivalent hyperplane. More specifically, for a given set of training examples $(\mathbf{x}_1, y_1), ..., (\mathbf{x}_n, y_n) \in \mathcal{H} \times \{-1, +1\}$ any hyperplane of the form (3.1) can be scaled such that

$$\min_{i=1,...,n} |\langle \mathbf{w}, \mathbf{x}_i \rangle + b| = 1$$

This is called a *canonical* hyperplane. The *margin* of a hyperplane is defined as the distance between the hyperplane and the point closest to it, i.e. for canonical hyperplanes the margin equals $1/\|\mathbf{w}\|$ (c.f. Schölkopf and Smola, 2002).

An important point is that one can prove that the capacity (e.g. the VC dimension) of the class of separating hyperplanes with a given margin decreases with increasing margin (Vapnik, 1979):

Theorem 3.2.1 (Vapnik, 1979). *Consider hyperplanes $\langle \mathbf{w}, \mathbf{x} \rangle = 0$, where $\mathbf{w}$ is normalized such that they are in canonical form w.r.t. a set of points $X^* = \{\mathbf{x}_1, ..., \mathbf{x}_r\}$. The set of decision functions $f_{\mathbf{w}}(\mathbf{x}) = \operatorname{sgn}(\langle \mathbf{w}, \mathbf{x} \rangle)$ defined on X^*, and satisfiying the constraint $||\mathbf{w}|| \leq \Lambda$, has a VC dimension satisfying $h \leq R^2\Lambda^2$. Here R is the radius of the smallest sphere centered at the origin and containing X^*.*

Looking back at the last Chapter that means the bigger the margin the lower becomes the right hand side of the VC bound (2.7)[1]. Hence, by making the margin as large as possible a bound on the risk is minimized. Theorem 3.2.1 therefore motivates to find the so-called *optimal* hyperplane (Boser et al., 1992; Cortes and Vapnik, 1995), which separates the two classes with maximum margin (Fig. 3.1). It can be constructed by

[1]Note that strictly speaking the theorem only applies, if we have chosen Λ a-priori.

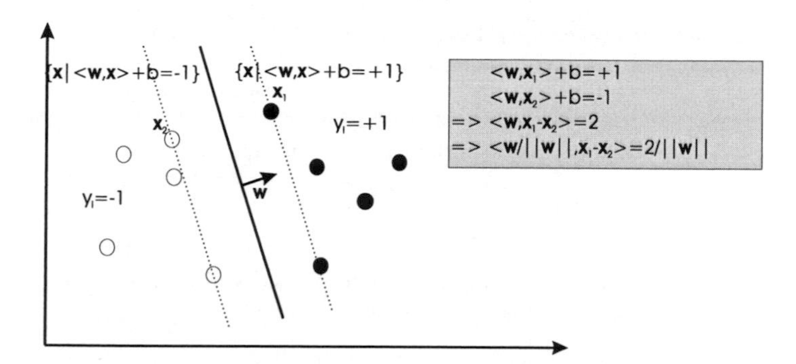

Figure 3.1: Optimal margin hyperplane in a binary classification problem: separate white dots from black dots. The *optimal* hyperplane is shown as a solid line. As the problem is linearly separable, there exists a weight vector $\mathbf{w}$ and a threshold b such that $y_i\left(\langle \mathbf{w}, \mathbf{x}_i \rangle + b\right) > 0$ $(i = 1, ..., n)$. Rescaling $\mathbf{w}$ and b such that the point(s) closest to the hyperplane satisfy $|\langle \mathbf{w}, \mathbf{x}_i \rangle + b| = 1$, we obtain a canonical form of the hyperplane satisfying $y_i\left(\langle \mathbf{w}, \mathbf{x}_i \rangle + b\right) \geq 1$. That means the margin in this case equals $1/\|\mathbf{w}\|$. This can be seen by considering two points $\mathbf{x}_1, \mathbf{x}_2$ on opposite sides of the hyperplane, which exactly satisfy $|\langle \mathbf{w}, \mathbf{x}_i \rangle + b| = 1$, and projecting them onto the hyperplane normal vector $\mathbf{w}/\|\mathbf{w}\|$ (c.f. Schölkopf and Smola, 2002).

solving the quadratic programming optimization problem:

$$\min_{\mathbf{w} \in \mathcal{H}, b \in \mathbb{R}} \frac{1}{2}\|\mathbf{w}\|^2 \quad \text{subject to } y_i\left(\langle \mathbf{x}_i, \mathbf{w} \rangle + b\right) \geq 1 \text{ for all } i = 1, ..., n \tag{3.3}$$

This is a convex optimization problem and hence its solution is unique. By introducing Lagrangian multipliers $\alpha = (\alpha_1, ..., \alpha_n)^T$ one can rewrite this primal optimization problem in its dual form as

$$\max_{\alpha \in \mathbb{R}^n} \sum_{i=1}^{n} \alpha_i - \frac{1}{2} \sum_{i,j=1}^{n} \alpha_i \alpha_j y_i y_j \langle \mathbf{x}_i, \mathbf{x}_j \rangle \quad \text{subject to } \alpha_i \geq 0 \text{ for all } i = 1, ..., n, \text{ and } \sum_{i=1}^{n} \alpha_i y_i = 0 \tag{3.4}$$

Due to the convexity of the primal problem Wolfe's theorem (Wolfe, 1961) guarantees the *Karush-Kuhn-Tucker* (KKT) duality-gap (Kuhn and Tucker, 1951) between the solution of the primal optimization problem and the dual optimization problem to be exactly zero, i.e. in both cases we obtain exactly the same hyperplane. Furthermore, the

KKT conditions imply that only the Lagrangian multipliers α_i, which are non-zero at the saddle point of (3.4), correspond to an exact equality in the constraints (Kuhn and Tucker, 1951). The patterns $\mathbf{x}_i$ associated to these Lagrangian multipliers are called *support vectors*. All support vectors are lying exactly on the margin, since only they can fulfill $y_i\left(\langle \mathbf{x}_i, \mathbf{w}\rangle + b\right) = 1$. Consequently, in the solution of the primal optimization problem

$$\mathbf{w} = \sum_{i=1}^{n} \alpha_i y_i \mathbf{x}_i \tag{3.5}$$

$\mathbf{w}$ only depends on the support vectors, i.e. on the patterns $\mathbf{x}_i$ with $\alpha_i \neq 0$. All remaining examples are irrelevant for the decision function

$$f_\alpha(\mathbf{x}) = \operatorname{sgn}\left(\sum_{i=1}^{n} \alpha_i y_i \langle \mathbf{x}, \mathbf{x}_i\rangle + b\right) \tag{3.6}$$

This nicely captures our intuition of the problem: As the hyperplane is completely determined by the training points closest to it, the solution does not depend on other examples.

Using the results in the previous Chapter it is straight forward to replace the dot product in (3.4) by a kernel function, which allows us to induce a nonlinear decision boundary in the original input space. At the same time in feature space classes +1 and -1 are still linearly separated by a hyperplane. The usage of kernels leads to the following modified formulation of the dual optimization problem (Cortes and Vapnik, 1995):

$$\max_{\alpha \in \mathbb{R}^n} \sum_{i=1}^{n} \alpha_i - \frac{1}{2} \sum_{i,j=1}^{n} \alpha_i \alpha_j y_i y_j k(x_i, x_j) \tag{3.7}$$

$$\text{subject to } \alpha_i \geq 0 \text{ for all } i = 1, ..., n, \text{ and } \sum_{i=1}^{n} \alpha_i y_i = 0$$

which induces a decision function

$$f_\alpha(x) = \operatorname{sgn}\left(\sum_{i=1}^{n} \alpha_i y_i k(x, x_i) + b\right) \tag{3.8}$$

This is the so-called *hard margin Support Vector Machine* (see Fig. 3.2). Note that our optimization problem is now formulated in the original input space $\mathcal{X}$, which is not necessarily a vector space anymore and thus gives us the freedom of dealing with arbitrarily structured objects.

3.2.2 Soft Margin Support Vector Machines

The C-SVM formulation Until now we focused on the case were no training errors were allowed and the data was perfectly separable by an optimal hyperplane in feature

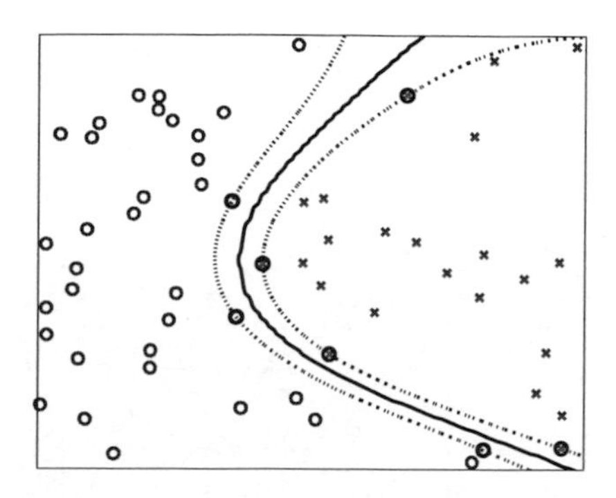

Figure 3.2: Example of a hard margin SVM using a RBF kernel. Crosses an circles are two classes of training examples. The solid line show the decision boundary within the margin (dotted). The SVs (marked with extra circles) are lying exactly on the margin.

space. However, in real world data this is rarely the case and thus a separating hyperplane does not exist. Cortes and Vapnik (1995) solve this problem by introducing so-called slack variables $\xi_i \geq 0,\ i = 1, ..., n$ and relaxing the separation constraints in (3.3) by just requiring

$$y_i \left(\langle \mathbf{x}_i, \mathbf{w} \rangle + b \right) \geq 1 - \xi_i,\ i = 1, ..., n \tag{3.9}$$

This is commonly referred to as a *soft margin hyperplane*. Obviously, these constraints could always be met by just making all ξ_i large enough. To compensate for that, the objective function (3.3) is modified as

$$\min_{\mathbf{w} \in \mathcal{H}, b \in \mathbb{R}} \frac{1}{2} \|\mathbf{w}\|^2 + C \sum_{i=1}^{n} \xi_i \tag{3.10}$$

where the constant $C > 0$ determines the trade-off between margin maximization and error minimization. Incorporating a kernel and the dual problem formulation, one has to maximize (3.7) subject to the constraints

$$0 \leq \alpha_i \leq C \text{ for all } i = 1, ..., n, \text{ and } \sum_{i=1}^{n} \alpha_i y_i = 0 \tag{3.11}$$

The only difference from the separable case is the upper bound C on the Lagrangian multipliers α_i. This way the influence of the individual patterns (which could be outliers) is limited. The solution again takes the same form as in (3.8). This is the so called *C-SVM*,

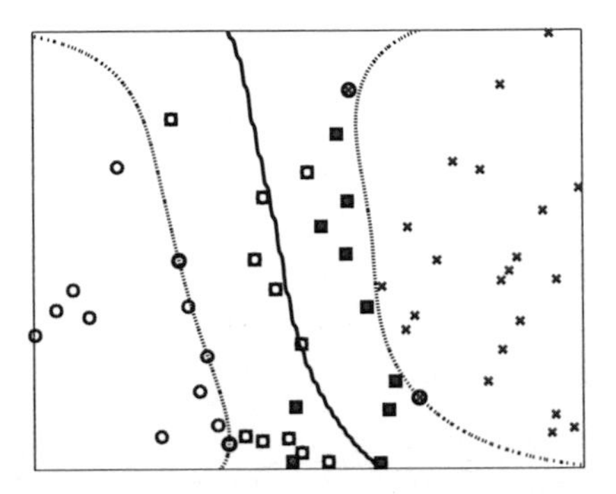

Figure 3.3: Example of a soft margin SVM: Crosses and circles are two classes of training examples. The solid line shows the decision boundary within the margin (dotted). The in-bound SVs (marked with extra circles) are lying exactly on the margin, while the rest of the support vectors become margin errors (marked with extra boxes).

which is the most commonly used SVM. Note that for $C \to \infty$ the solution converges to the solution of the hard margin SVM. However, in contrast to the hard margin case, for the soft margin SVM not all support vectors lie exactly on the margin. Support vectors with $\alpha_i = C$ (*bound support vectors*) usually, but not always, lie *within* the margin, i.e. they correspond to slack variables $\xi_i > 0$. They are then also called *margin errors*. In contrast, support vectors with $0 < \alpha_i < C$ (*in-bound support vectors*) lie always exactly on the margin (see Fig. 3.3).

There is an important connection of the soft margin SVM formulation to the Tikhonov regularization framework described in the last Chapter. Let g_α be the decision function (3.8) without the sign. Ignoring for the sake of simplicity the offset b and denoting $\tilde{\alpha}_i = y_i \alpha_i$ we can rewrite $g_\alpha(x)$ as $g_{\tilde{\alpha}}(x) = \sum_{i=1}^{n} \tilde{\alpha}_i k(x, x_i)$. We thus recover a function of the form (2.9) in a RKHS $\mathcal{F}$ induced by the kernel k. Moreover, we have

$$\|g_{\tilde{\alpha}}\|_{\mathcal{F}}^2 = \left\langle \sum_{i=1}^{n} \tilde{\alpha}_i k(x, x_i), \sum_{i=1}^{n} \tilde{\alpha}_i k(x, x_i) \right\rangle_{\mathcal{F}} = \sum_{i,j=1}^{n} \tilde{\alpha}_i \tilde{\alpha}_j k(x_i, x_j) = \|\mathbf{w}\|^2 \tag{3.12}$$

This shows that within the Tikhonov regularization framework $\|\mathbf{w}\|^2$ can be understood as a special regularizer maximizing the smoothness of the decision function f in input domain $\mathcal{X}$. The regularization parameter is set to $\lambda = \frac{1}{C}$ and the loss function is

$$\ell(y, \hat{y}) = \max(0, 1 - y\hat{y}) \tag{3.13}$$

which is also called *hinge loss* (c.f. Schölkopf and Smola, 2002).

Estimating the Generalization Capability In contrast to the hard margin SVM the generalization capability of the soft margin SVM no longer depends on the size of the margin only, but on the trade-off between the fraction of points with $\xi_i > 0$ (margin errors) and the size of the margin (Bartlett and Shawe-Taylor, 1999):

Theorem 3.2.2 (Margin Error Bound, Bartlett and Shawe-Taylor, 1999). *Consider the set of descision functions $f(\mathbf{x}) = \operatorname{sgn}\langle \mathbf{w}, \mathbf{x}\rangle$ with $\|\mathbf{w}\| \leq \Lambda$ and $\|\mathbf{x}\| \leq R$ for some $R, \Lambda > 0$. Moreover, let $\rho > 0$, and ν denote the fraction of training examples with $\xi_i > 0$ (margin errors).*

For all distributions $\mathcal{P}$ generating the data, with probability of at least $1 - \delta$ over the drawing of the n training patterns, and for any $\rho > 0$ and $\delta \in (0, 1)$, the probability that a test pattern $(\hat{x}, \hat{y})$ drawn from $\mathcal{P}$ will be misclassified is bounded from above by

$$P(f_\alpha(\hat{x}) \neq \hat{y}) \leq \nu + \sqrt{\frac{\lambda}{n}\left(\frac{R^2\Lambda^2}{\rho^2}\log^2 n + \log(1/\delta)\right)} \tag{3.14}$$

where λ is some constant.

Note that in case of the C-SVM we set $\rho = 1$ by using canonical hyperplanes and $\Lambda = \|\mathbf{w}\|$. Similarly to (3.14) the bound also includes the radius of the smallest sphere containing the data in feature space. It can be seen as a normalization factor, because the size of the margin clearly depends on the kernel function and hence on the feature space we are working in. It can be computed as the solution of the following quadratic program (Schölkopf et al., 1995):

$$\max_{\beta} \sum_{i=1}^{n} \beta_i k(x_i, x_i) - \sum_{i,j=1}^{n} \beta_i \beta_j k(x_i, x_j) \tag{3.15}$$

$$\text{subject to} \sum_{i=1}^{n} \beta_i = 1, \beta_i \geq 0$$

Another important estimation of the generalization capability was given by Vapnik and Chapelle (2000). They use a quantity called the *span* of the support vectors, to approximate the leave-one-out cross-validation error:

Definition 3.2.3 (Span of SV solution). W.l.o.g. assume $\alpha_1, ..., \alpha_m$ to be the Lagrangian multipliers associated to the **in-bound** support vectors $x_1, ..., x_m$. Then the span S_p of support vector x_p is defined as the minimal distance between x_p and the set

$$\Lambda_p := \left\{\lambda \middle| \sum \lambda_i = 1, \text{ and } 0 \leq \alpha_i + y_i y_p \alpha_p \lambda_i \leq C\right\} \tag{3.16}$$

Chapelle et al. (2002) show that the value of the span can be computed in closed form as

$$S_p^2 = \frac{1}{\left(\tilde{\mathbf{K}}_{SV}^{-1}\right)_{pp}} \tag{3.17}$$

where

$$\tilde{\mathbf{K}}_{SV} = \begin{pmatrix} \mathbf{K}_{SV} & \mathbf{1} \\ \mathbf{1}^T & 0 \end{pmatrix} \tag{3.18}$$

and $\mathbf{K}_{SV}$ is the kernel matrix restricted to support vectors and $\mathbf{1}$ denotes a vector of ones. Furthermore, Vapnik and Chapelle (2000) prove the following approximation for the leave-one-out error:

Theorem 3.2.4 (Span rule, Vapnik and Chapelle, 2000). *If the set of support vectors remains the same during the leave-one-out procedure, the leave-one-out error is given as*

$$R_{LOO} = \frac{1}{n} \left| \left\{ p \,\middle|\, \alpha_p S_p^2 \geq y_p f_\alpha(x_p) \right\} \right| \tag{3.19}$$

Of course, the assumption of this theorem is rarely exactly fulfilled in practice, but the number of violations is usually small and so the span rule is usually regarded as a quite accurate method to estimate the generalization performance of a SVM efficiently (Vapnik and Chapelle, 2000).

Besides the two estimations of the generalization error cited here there exist some other bounds (e.g. Joachims, 2000; Luxburg et al., 2004), which are, however, not of importance for this thesis.

The ν-SVM formulation Another popular formulation for soft margin SVMs is the so-called *ν-SVM* (Schölkopf et al., 2000). Instead of using the parameter C, one uses a parameter $\nu \in [0, 1]$ which can be shown to provide a lower bound on the fraction of examples that will be support vectors and upper bounds for the fraction of margin errors (Schölkopf et al., 2000). The primal optimization problem for the ν-SVM is formulated as

$$\max_{\mathbf{w} \in \mathbb{R}, \xi \in \mathbb{R}^n, \rho, b \in \mathbb{R}} \frac{1}{2} \|\mathbf{w}\|^2 - \nu\rho + \sum_{i=1}^{n} \xi_i \tag{3.20}$$

$$\text{subject to } y_i \left(\langle \mathbf{x}_i, \mathbf{w} \rangle + b \right) \geq \rho - \xi_i \text{ , and } \xi_i \geq 0 \text{ for all } i = 1, ..., n, \ \rho \geq 0$$

where ρ is a parameter that controls the size of the margin[2]. The dual problem in this case can be derived as

$$\max_{\alpha} -\frac{1}{2}\sum_{i,j=1}^{n} \alpha_i \alpha_j y_i y_j k(x_i, x_j) \tag{3.21}$$

$$\text{subject to } 0 \leq \alpha_i \leq 1, \ \sum_{i=1}^{n} \alpha_i y_i = 0, \ \sum_{i=1}^{n} \alpha_i \geq \nu$$

The resulting decision function has the same form as in the C-SVM case. It can be shown that, if the ν-SVM leads to $\rho > 0$, the decision function is identical to the one which can be obtained by a C-SVM with $C = 1/\rho$ (Schölkopf et al., 2000).

Multi-class SVMs Until now we have only focused on the two-class case. If we have more than just two classes to separate, say M, common strategies are the following (see e.g. Schölkopf and Smola, 2002):

- Construct a set of M binary classifiers $f^{(1)}, ..., f^{(M)}$, each trained to separate one class from the rest (*one-against-rest* approach). The cth classifier computes a decision boundary between class c and the other $M-1$ classes. An unseen test example x is assigned to the class for which the distance of the image $\phi(x)$ from the hyperplane in the positive direction (i.e. in the direction of class c) is maximal.
- Construct a set of $(M-1)M/2$ binary classifiers $f^{(1,2)}, f^{(1,3)} ..., f^{(M-1,M)}$, each trained to separate one pair of classes (*one-versus-one* approach). A classifier $f^{(c,\tilde{c})}$ separating class c from class $\tilde{c}$ returns +1, if the image $\phi(x)$ of an unseen test instance x has a positive distance from the hyperplane and -1 otherwise. A return of +1 can be interpreted as a vote for class c, while a return of -1 can be interpreted as a vote for class $\tilde{c}$. The example x is assigned to the class, which gets the maximum number of votes of all classifiers.
  At the first glance this seems to be less effective than the one-against-rest approach, since there are more binary classifiers to train. On the other hand, the advantage of this method is that the problems to be learned are usually simpler (because only the patterns belonging to two classes have to be taken into account), and they are not as imbalanced as in the other approach. Especially with a high number of classes one usually would get a high number of negative versus a small number of positive examples, which can cause problems for the SVM to obtain a good generalization performance.

Other techniques include error-correcting output coding and directed acyclic graphs (c.f. Schölkopf and Smola, 2002). Besides, there exists also a multi-class version of the SVM

[2]Note that ρ in this case is *not* necessarily 1, since ν-SVMs do not use canonical hyperplanes.

(Weston and Watkins, 1999). Further extensions and variations of SVM classification can be found in the books by Schölkopf and Smola (2002) and Shawe-Taylor and Cristianini (2004).

3.3 Kernel Ridge Regression

We now turn to the problem of learning a real-valued function $f : \mathcal{X} \to \mathbb{R}$ via kernel-based machine learning techniques. That means we have training data $\mathcal{D} = \{(x_i, y_i)\}_{i=1}^n \subset \mathcal{X} \times \mathbb{R}$ drawn i.i.d. from some unknown probability distribution $\mathcal{P}(x, y)$. *Ridge regression* is a classical algorithm for estimating a linear regression function in a robust way (e.g. Duda et al., 2001). In the following we consider a kernelized version of this method (c.f. Shawe-Taylor and Cristianini, 2004): Assuming the squared loss function

$$\ell(y, \hat{y}) = (y - \hat{y})^2 \tag{3.22}$$

one seeks for a regression function f that is statistically stable and robust. Following Chapter 2.4 we can consider f to be a function in a RKHS $\mathcal{F}$ and perform regularized risk minimization with the regularization term $\|f\|_{\mathcal{F}}^2$. That means we are looking for a regression function with maximal smoothness on one hand and a good fit of the training data on the other hand. Due to the representer theorem the solution f_α has the kernel expansion

$$f_\alpha(x) = \sum_{i=1}^{n} \alpha_i k(x, x_i) \tag{3.23}$$

Equivalently, we can think of f_α as a linear function in a feature space $\mathcal{H}$ of the form $f_{\mathbf{w}} = \langle \mathbf{w}, \phi(x) \rangle$ (c.f. Chapter 2.5). Consequently, similar to the SVM case, we have $\mathbf{w} = \sum_i \alpha_i \phi(x_i)$ and $\|\mathbf{w}\|^2 = \|f_\alpha\|_{\mathcal{F}}^2$. This justifies minimizing the trade-off between the norm of the weight vector on one hand and the sum squared error on the other hand (c.f. Shawe-Taylor and Cristianini, 2004):

$$\min_{\mathbf{w}} \sum_{i=1}^{n} \xi_i^2 + \tau \|\mathbf{w}\|^2 \tag{3.24}$$
$$\text{subject to } y_i - \langle \mathbf{w}, \phi(x_i) \rangle = \xi_i, i = 1, ..., n$$

This is a convex optimization problem, and similar to the last Section we can rewrite it in its dual form with the Lagrangian multipliers $\alpha = (\alpha_1, ..., \alpha_n)^T$:

$$\max_{\alpha} \tau \sum_{i=1}^{n} \alpha_i^2 - 2 \sum_{i=1}^{n} \alpha_i y_i + \sum_{i,j=1}^{n} \alpha_i \alpha_j k(x_i, x_j) \tag{3.25}$$

which leads to the solution

$$\alpha = (\mathbf{K} + \tau \mathbf{I})^{-1} \mathbf{y}$$

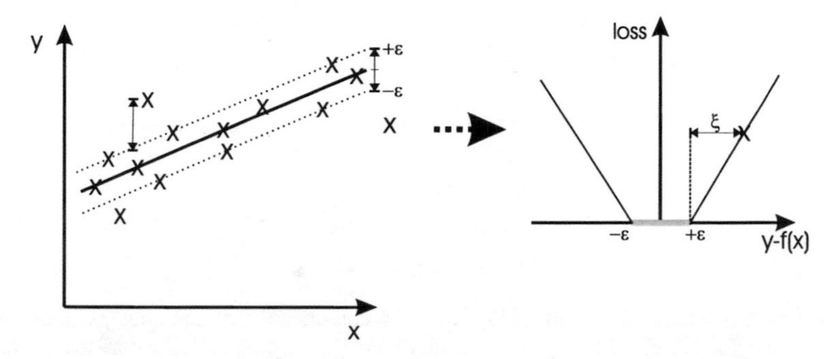

Figure 3.4: Only training points that lie outside the ϵ-tube are penalized in a linear fashion (c.f. Schölkopf and Smola, 2002).

with $\mathbf{y} = (y_1, ..., y_n)^T$and $\mathbf{I}$ being the identity matrix. Thereby the *ridge parameter* τ plays the role of a regularization constant.

The advantage of kernel ridge regression is the simplicity of the algorithm. The disadvantage is that in contrast to SVMs the solution is **not** sparse, i.e. in general we have as many $\alpha_i \neq 0$ as we have training data. This makes this approach slow for testing.

3.4 Support Vector Machines for Regression

SVMs for regression, also called Support Vector Regression (SVR), were mainly developed to introduce the benefit of sparsity into kernel-based regression estimation. For this purpose Vapnik (1995) devised the so-called *ϵ-insensitive loss function*

$$\ell(y, \hat{y}) = |y - \hat{y}|_\epsilon := \max\{0, |y - \hat{y}| - \epsilon\} \tag{3.26}$$

The idea is that only training points lying outside a small ϵ-tube around the estimated function contribute to the training error and are penalized in a linear fashion (Fig. 3.4).

Similarly to Kernel Ridge Regression, one seeks for a regression function $f_\alpha(x) = \sum_{i=1}^n \alpha_i k(x, x_i) + b$ with small norm $\|f_\alpha\|_\mathcal{F}$, i.e. a function that is as smooth as possible on one hand and on the other hand leads to a low empirical risk. As above this corresponds to constructing a linear regression function $f_\mathbf{w}(x) = \langle \mathbf{w}, \phi(x) \rangle + b$ in feature space $\mathcal{H}$ with small norm $\|\mathbf{w}\|^2 = \|f_\alpha\|^2_\mathcal{F}$. Introducing slack variables ξ_i, ξ_i^* to penalize points that are above or below the ϵ-tube the underlying optimization problem can be formulated as a quadratic program (Vapnik, 1995):

$$\min_{\mathbf{w}\in\mathcal{H},b\in\mathbb{R},\xi^{(*)}\in\mathbb{R}^n} \frac{1}{2}\|\mathbf{w}\|^2 + C\sum_{i=1}^{n}(\xi_i + \xi_i^*) \tag{3.27}$$

$$\begin{aligned} &\text{subject to:} \\ &y_i - \langle \mathbf{w}, \phi(x_i)\rangle + b \leq \epsilon + \xi_i \\ &\langle \mathbf{w}, \phi(x_i)\rangle + b - y_i \leq \epsilon + \xi_i^* \\ &\xi_i^{(*)} \geq 0 \end{aligned}$$

It is worth mentioning that in analogy to SVM classification minimizing $\|\mathbf{w}\|^2$ can also be understood as maximizing the so-called *ϵ-margin*, which is defined as the minimal distance between two patterns $\phi(x), \phi(x')$ in feature space with $|f_{\mathbf{w}}(x) - f_{\mathbf{w}}(x')| \geq 2\epsilon$. Note that this definition includes the definition of the margin given in Section 3.2 for the classification case: Especially for canonical hyperplanes we have $\epsilon = 1$. In general, following the derivation in Section 3.2, the size of the ϵ-margin equals $\epsilon/\|\mathbf{w}\|$. Maximizing the ϵ-margin therefore corresponds to finding a *flat* hyperplane in feature space, which in turn is equivalent to maximizing the smoothness of the function in input space.

Analogously to SVM classification, the primal optimization problem (3.27) is a convex problem and can thus equivalently be formulated in its dual form by introducing Lagrangian multipliers $\alpha = (\alpha_1, ..., \alpha_n)^T$ and $\alpha^* = (\alpha_1^*, ..., \alpha_n^*)^T$, which allows for using kernels:

$$\begin{aligned} \max_{\alpha} -\frac{1}{2}\sum_{i,j=1}^{n}(\alpha_i - \alpha_i^*)(\alpha_j - \alpha_j^*)k(x_i, x_j) \\ -\epsilon\sum_{i=1}^{n}(\alpha_i + \alpha_i^*) + \sum_{i=1}^{n} y_i(\alpha_i - \alpha_i^*) \\ \text{subject to} \sum_{i=1}^{n}(\alpha_i - \alpha_i^*) = 0 \text{ and } 0 \leq \alpha_i^{(*)} \leq C \end{aligned} \tag{3.28}$$

The resulting weight vector $\mathbf{w}$ can be written as

$$\mathbf{w} = \sum_{i=1}^{n}(\alpha_i - \alpha_i^*)\phi(x) \tag{3.29}$$

and hence the regression function has the form

$$f_\alpha(x) = \sum_{i=1}^{n}(\alpha_i - \alpha_i^*)k(x_i, x) + b \tag{3.30}$$

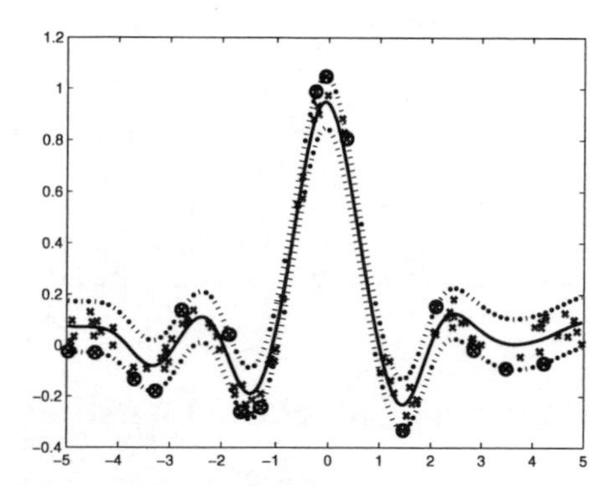

Figure 3.5: Support Vector Regression: The regression function (solid line) wrapped by the ϵ-tube (dotted) only depends on the SVs, which lie outside or exactly on the ϵ-insensitive tube (marked by extra circles).

Like in the SVM classification case, according to the KKT conditions (c.f. Section 3.2.1), only Lagrangian multipliers α_i, α_i^* which are non-zero at the saddle point of (3.28) correspond to an exact equality in the constraints and are thus associated to support vectors. These are only points with $|f_{\mathbf{w}}(x_i) - y_i| \geq \epsilon$. Thus the regression function (3.30) only depends on the points lying outside or exactly on the ϵ-insensitive tube (Fig. 3.5).

The use of the ϵ-insensitive loss function has an impact on the probabilistic model, which is inherently employed in SVR to model the noise in the y-measurements: While in Kernel Ridge Regression measurements are assumed to be disturbed by some Gaussian distributed noise, in SVR we have a noise density of the form $\frac{1}{2(1+\epsilon)} \exp(-|(y - \hat{y}|_\epsilon)$ (Smola and Schölkopf, 1998).

Concerning the generalization performance of a SVR, Shawe-Taylor and Cristianini (2004) report an upper bound on the expectation that the difference between the true function value y and the estimated function value $\hat{y}$ at some point x exceeds a threshold γ:

Theorem 3.4.1 (Shawe-Taylor and Cristianini). *Let R be the smallest sphere enclosing the data in feature space. With probability at least $1 - \delta$ over the random draw of the dataset $\mathcal{D}$, we have*

$$\mathbf{E}[\delta(|y - \hat{y}| > \gamma)] \leq \frac{1}{n(\gamma - \epsilon)} \sum_i (\xi_i + \xi_i^*) + 4\sqrt{\frac{R^2\|\mathbf{w}\|^2}{n(\gamma - \epsilon)^2}} + 3\sqrt{\frac{\log(2/\delta)}{2n}} \tag{3.31}$$

Note the similarity of this bound to the radius-margin error bound (Eq.: 3.14) in SVM classification.

It should be noted that there also exists a formulation of the SVR using the ν-trick as described for SVM classification. Therefore, it is usually distinguished between ϵ-SVR and ν-SVR. For this thesis only the ϵ-SVR is of relevance. Further extensions and variations of SVR can be found in the books by Schölkopf and Smola (2002) and Shawe-Taylor and Cristianini (2004). An efficient C++ implementation of SVM classification and SVR, which we utilized in this thesis, can be found in the LIBSVM software (Chang and Lin, 2001).

3.5 Kernel Canonical Correlation Analysis

Canonical Correlation Analysis (CCA — Hotelling, 1936) is a classical method for extracting common features from two different views of the data in an unsupervised manner. That means, we consider a dataset $\mathcal{D} = \{x_i\}_{i=1}^n \subset \mathcal{X}$, in which each object x_i has a representation $\phi_a(x_i)$ in some feature space $\mathcal{H}_a$ and at the same time there is another representation $\phi_b(x_i)$ in a different feature space $\mathcal{H}_b$. As an example consider a representation of a molecule as a graph on one hand and a representation by certain descriptors on the other hand. (This will play a role later on in Chapter 6.) Clearly, both feature spaces are views of the same molecular structures, but the mathematical description is different. By seeking for correlations between the $\{\phi_a(x_i)\}_{i=1}^n$ and $\{\phi_b(x_i)\}_{i=1}^n$, one might hope to extract features that bring out the relevant chemical properties.

Practically this is done by rotating the coordinate systems in both feature spaces, such that the newly constructed axes successively extract directions of maximal correlation between both views in a decreasing order. So there is a certain similarity to PCA (Hotelling, 1933): While in PCA the principal components extract directions of maximal variance within one view, in CCA the canonical correlates are the directions of maximal correlation between two different views. By projecting the data on the canonical correlates common features from both representations are extracted.

Mathematically this is achieved by considering projections of the form $f_a^t(x_i) = \langle \mathbf{w}_a^t, \phi_a(x_i)\rangle$ and $f_b^t(x_i) = \langle \mathbf{w}_b^t, \phi_b(x_i)\rangle$ in feature spaces $\mathcal{H}_a$ and $\mathcal{H}_b$, respectively. Here $f_a^t(x_i)$ and $f_b^t(x_i)$ denote the projections of object x_i on the tth canonical correlates $\mathbf{w}_a^t$ and $\mathbf{w}_b^t$. Now, the goal is to maximize the empirical correlation between the $f_a^t(x_i)$ and $f_b^t(x_i)$ with respect to $\mathbf{w}_a^t, \mathbf{w}_b^t$:

$$\begin{aligned} \max_{\mathbf{w}_a^t, \mathbf{w}_b^t} \quad & \frac{\widehat{\mathbf{Cov}}\left[\langle \mathbf{w}_a^t, \phi_a(x)\rangle, \langle \mathbf{w}_b^t, \phi_b(x)\rangle\right]}{\sqrt{\widehat{\mathbf{Var}}\left[\langle \mathbf{w}_a^t, \phi_a(x)\rangle\right] \widehat{\mathbf{Var}}\left[\langle \mathbf{w}_b^t, \phi_b(x)\rangle\right]}} \\ = \quad & \frac{(\mathbf{w}_a^t)^T \mathbf{C}_{ab} \mathbf{w}_b^t}{\sqrt{(\mathbf{w}_a^t)^T \mathbf{C}_{aa} \mathbf{w}_a^t (\mathbf{w}_b^t)^T \mathbf{C}_{bb} \mathbf{w}_b^t}} \end{aligned} \tag{3.32}$$

Here $\widehat{\mathbf{Var}}$ and $\widehat{\mathbf{Cov}}$ denote the empirical variance and covariance, respectively. $\mathbf{C}_{aa}, \mathbf{C}_{bb}$ are the empirical covariance matrices for objects within $\mathcal{H}_a$ and $\mathcal{H}_b$, and $\mathbf{C}_{ab}$ is the empirical covariance matrix between objects embedded in $\mathcal{H}_a$ and those embedded in $\mathcal{H}_b$. With the additional normalization $\widehat{\mathbf{Var}}\left[\langle \mathbf{w}_a^t, \phi_a(x)\rangle\right] = 1$ and $\widehat{\mathbf{Var}}\left[\langle \mathbf{w}_b^t, \phi_b(x)\rangle\right] = 1$ the solution of (3.32) can be found by solving the generalized eigenvector problem

$$\begin{pmatrix} 0 & \mathbf{C}_{ab} \\ \mathbf{C}_{ba} & 0 \end{pmatrix} \begin{pmatrix} \mathbf{w}_a^t \\ \mathbf{w}_b^t \end{pmatrix} = \rho_t \begin{pmatrix} \mathbf{C}_{aa} & 0 \\ 0 & \mathbf{C}_{bb} \end{pmatrix} \begin{pmatrix} \mathbf{w}_a^t \\ \mathbf{w}_b^t \end{pmatrix} \tag{3.33}$$

Thereby the eigenvalue ρ_t gives the size of the correlation between the $f_a^t(x_i)$ and $f_b^t(x_i)$.

It is possible to reformulate problem (3.33) in its dual form to allow the introduction of kernel functions (c.f. Lai and Fyfe, 2000; Akaho, 2001; Bach and Jordan, 2002): Note that in analogy to PCA f_a^t and f_b^t can be expressed as linear combinations of the ϕ-images of the training data. Thus we can write $\mathbf{w}_a^t = \sum_{j=1}^n \alpha_j^t \phi_a(x_j)$, $\mathbf{w}_b^t = \sum_{j=1}^n \beta_j^t \phi_b(x_j)$. Hence, we have $f_a^t(x_i) = \sum_{j=1}^n \alpha_j^t k_a(x_i, x_j)$ and $f_b^t(x_i) = \sum_{j=1}^n \beta_j^t k_b(x_i, x_j)$, where k_a and k_b are the kernel functions corresponding to both feature spaces. Therefore, the projections on the canonical correlates can be interpreted as functions $f_a^t : \mathcal{X} \to \mathbb{R}$, $f_b^t : \mathcal{X} \to \mathbb{R}$ in the input space. They allow the extraction of correlated features from $\mathcal{H}_a$ and $\mathcal{H}_b$ via kernel functions k_a and k_b. These features are always real valued an thus effectively transform an arbitrarily structured domain into a vector representation. Moreover, f_a^t and f_b^t can be viewed as functions in a RKHS $\mathcal{F}$. The goal is now to find dual variables $\alpha^t = (\alpha_1^t, ..., \alpha_n^t)^T$ and $\beta^t = (\beta_1^t, ..., \beta_n^t)^T$ such that the empirical correlation between f_a^t and f_b^t is maximized:

$$\begin{aligned} \max_{\alpha^t, \beta^t} \quad & \frac{\widehat{\mathbf{Cov}}\left[f_a^t(x), f_b^t(x)\right]}{\sqrt{\widehat{\mathbf{Var}}\left[f_a^t(x)\right] \widehat{\mathbf{Var}}\left[f_b^t(x)\right]}} \\ = \quad & \frac{(\alpha^t)^T \mathbf{K}_a \mathbf{K}_b \beta^t}{\sqrt{((\alpha^t)^T \mathbf{K}_a^2 \alpha^t)^T \, ((\beta^t)^T \mathbf{K}_b^2 \beta^t)}} \end{aligned} \tag{3.34}$$

Here $\mathbf{K}_a$ and $\mathbf{K}_b$ denote the kernel matrices associated to both feature spaces.

The problem with maximizing (3.34) is that the estimated empirical correlations might not be statistically stable and robust, i.e. f_a^t and f_b^t might not generalize well. Therefore, the Tikhonov regularization framework is employed to enforce the smoothness of f_a^t and f_b^t by requiring small squared norms $\|f_a^t\|_{\mathcal{F}}^2 = \|\mathbf{w}_a^t\|^2 = (\alpha^t)^T \mathbf{K}_a \alpha^t$ and $\|f_b^t\|_{\mathcal{F}}^2 =$

$\|\mathbf{w}_b^t\|^2 = (\beta^t)^T \mathbf{K}_b \beta^t$. In conclusion we arrive at the modified optimization problem

$$\max_{\alpha^t,\beta^t} \quad \frac{\widehat{\mathbf{Cov}}\left[f_a^t(x), f_b^t(x)\right]}{\sqrt{\left(\widehat{\mathbf{Var}}\left[f_a^t(x)\right] + \tau\|f_a^t\|_{\mathcal{F}}^2\right)\left(\widehat{\mathbf{Var}}\left[f_b^t(x)\right] + \tau\|f_b^t\|_{\mathcal{F}}^2\right)}} \tag{3.35}$$

$$= \quad \frac{(\alpha^t)^T \mathbf{K}_a \mathbf{K}_b \beta^t}{\sqrt{((\alpha^t)^T \mathbf{K}_a^2 \alpha^t + \tau(\alpha^t)^T \mathbf{K}_a \alpha^t)^T \, ((\beta^t)^T \mathbf{K}_b^2 \beta^t + \tau(\beta^t)^T \mathbf{K}_b \beta^t)}}$$

where τ is a regularization constant. If we approximate $(\alpha^t)^T \mathbf{K}_a^2 \alpha^t + \tau(\alpha^t)^T \mathbf{K}_a \alpha^t \approx (\alpha^t)^T (\mathbf{K}_a + \frac{\tau}{2}\mathbf{I})^2 \alpha^t$ ($\mathbf{I}$ is the identity matrix), we arrive at the generalized eigenvector problem

$$\begin{pmatrix} \mathbf{0} & \mathbf{K}_a\mathbf{K}_b \\ \mathbf{K}_b\mathbf{K}_a & \mathbf{0} \end{pmatrix} \begin{pmatrix} \alpha^t \\ \beta^t \end{pmatrix} = \rho_t \begin{pmatrix} (\mathbf{K}_a + \frac{\tau}{2}\mathbf{I})^2 & \mathbf{0} \\ \mathbf{0} & (\mathbf{K}_b + \frac{\tau}{2}\mathbf{I})^2 \end{pmatrix} \begin{pmatrix} \alpha^t \\ \beta^t \end{pmatrix} \tag{3.36}$$

with $(\alpha^t, \beta^t)^T$ being the tth eigenvector corresponding to eigenvalue ρ_t. A projection of training and test data can then be computed by applying f_a^t and f_b^t.

It is also possible to extend KCCA to deal with more than two data representations. Further readings on CCA and KCCA can be found in the book by Shawe-Taylor and Cristianini (2004) and the papers by Lai and Fyfe (2000); Akaho (2001) and Bach and Jordan (2002).

3.6 Kernel-Based Clustering

3.6.1 Discovering Cluster Structure in Feature Space

Cluster analysis, in contrast to classification, is a completely unsupervised technique, which aims to discover separated groups of data items within the complete dataset (c.f. Jain et al., 1999; Duda et al., 2001). That means given data $\mathcal{D} = \{x_i\}_{i=1}^n \subset \mathcal{X}$ we would like to learn a function $f : \mathcal{X} \rightarrow \{1, ..., K\}$, which assigns each object x_i to its corresponding cluster $c \in \{1, ..., K\}$. As usual, we assume our original data to be mapped to some feature space $\mathcal{H}$ via the function $\phi : \mathcal{X} \rightarrow \mathcal{H}$.

The goal of a clustering algorithm is that objects assigned to the same cluster have a high within-cluster similarity and at the same time low similarity to all other objects not in the same cluster. By performing a clustering, the data is broken down into a number of groups $\mathcal{D}_1, ..., \mathcal{D}_K$ composed of objects, which are significantly more uniform than the overall dataset. This may help a human's understanding of the data.

Classically, clustering algorithms can for instance be categorized as hierarchical vs. partitional, agglomerative vs. divisive, hard vs. fuzzy and deterministic vs. stochastic/-probabilistic (Jain et al., 1999). Probably most prominent examples for clustering methods are the linkage algorithms, which are hierarchical agglomerative approaches, and the k-means algorithm, which is a partitional method (e.g. Jain et al., 1999). Additionally, spectral clustering techniques have attracted a high attention during the last years (Weiss, 1999; Shi and Malik, 2000; Meila and Shi, 2001a; Kannan and Vempala, 2000; Meila and Shi, 2001b; Ng et al., 2002; c.f. review paper by Verma and Meila, 2003). All these methods can be formulated in terms of kernel functions (c.f. Shawe-Taylor and Cristianini, 2004).

In the following we give a brief review over linkage algorithms, dual k-means and spectral clustering, which are the methods employed in this thesis. Further readings on clustering can be found e.g. in the book by Jain and Dubes (1988).

3.6.2 Linkage Algorithms

Linkage methods are hierarchical clustering approaches. At the beginning each data point forms one cluster. Then those clusters are merged, which have maximum similarity according to some criterion. This criterion can either be (c.f. Jain et al., 1999)

1. the maximal similarity (i.e. minimal distance) between any two points from a cluster pair. This leads to single linkage clustering.

2. the minimum similarity (i.e. maximal distance) between any two points from a cluster pair. This leads to complete linkage clustering.

3. the average similarity between any two points from a cluster pair. This leads to average linkage clustering.

Merging clusters is continued, until only one cluster is left. Through the merging process at the end clusters are organized in a tree-like fashion with the clusters containing only one data point at the bottom and the cluster containing all data points at the top of the tree. This can be visualized in form of a dendrogram (Fig. 3.6).

Obviously, the merging process in linkage clustering is guided by a similarity function between individual data points. As a kernel function can be viewed as a special similarity measure, the application to linkage clustering is natural.

3.6.3 Kernel/Dual k-means

The k-means clustering algorithm is a very well known partitioning method and thus we omit a detailed description here. The idea is to minimize for each point the sum of distances to its cluster centroid

$$\sum_{c=1}^{K} \sum_{\phi(x)\in\mathcal{D}_c} \|\phi(x) - \mu_c\|^2 = \sum_{i=1}^{n} \|\phi(x_i) - \mu_{f(x_i)}\|^2 \tag{3.37}$$

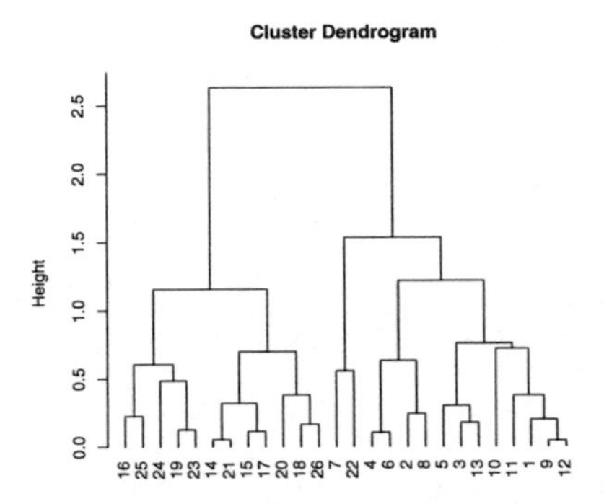

Figure 3.6: Example of a linkage clustering tree. The y-axis shows the dissimilarity between clusters and the x-axis their labels.

in a greedy hill-climbing fashion. Here μ_c is the cluster centroid of the cth cluster. It can be shown that the minimization of the sum-of-squared distances criterion at the same time corresponds to a maximization of the within-cluster similarity and the between-cluster dissimilarity and thus offers a good criterion for clustering (e.g. Shawe-Taylor and Cristianini, 2004). It leads to compact and isotropic clusters that are linearly separable. In contrast, single linkage, for instances, detects elongated, chain-like clusters (Jain et al., 1999).

It is relatively easy to rewrite expression (3.37) in terms of kernels (Shawe-Taylor and Cristianini, 2004): Let $\mathbf{A} \in \mathbb{R}^{n \times K}$ denote an indicator matrix with $A_{ic} = 1$ if pattern x_i belongs to cluster c and 0 otherwise. Let further be $\mathbf{D} \in \mathbb{R}^{K \times K}$ a diagonal matrix with $D_{cc} = \frac{1}{|\mathcal{D}_c|} = \frac{1}{\sum_i A_{ic}}$ and $\mathbf{k} = (k(x, x_1), ..., k(x, x_n))^T$ the vector of inner products between the image of some test pattern $\phi(x)$ and the training examples in feature space. Then with $\mu_c := \frac{1}{|\mathcal{D}_c|} \sum_{x' \in \mathcal{D}_c} \phi(x')$ it is

$$\begin{aligned} \|\phi(x) - \mu_c\|^2 &= \|\phi(x)\|^2 - 2\langle \phi(x), \mu_c \rangle + \|\mu_c\|^2 \\ &= k(x,x) - 2\frac{1}{|\mathcal{D}_c|} \sum_{x' \in \mathcal{D}_c} k(x, x') + \frac{1}{|\mathcal{D}_c|^2} \sum_{x', x'' \in \mathcal{D}_c} k(x', x'') \\ &= k(x,x) - 2(\mathbf{k}^T \mathbf{A} \mathbf{D})_c + (\mathbf{D} \mathbf{A}^T \mathbf{K} \mathbf{A} \mathbf{D})_{cc} \end{aligned} \tag{3.38}$$

Hence a data item x should be assigned to the cluster c for which $(\mathbf{D}\mathbf{A}^T\mathbf{K}\mathbf{A}\mathbf{D})_{cc} - 2(\mathbf{k}^T\mathbf{A}\mathbf{D})_c$ is minimal.

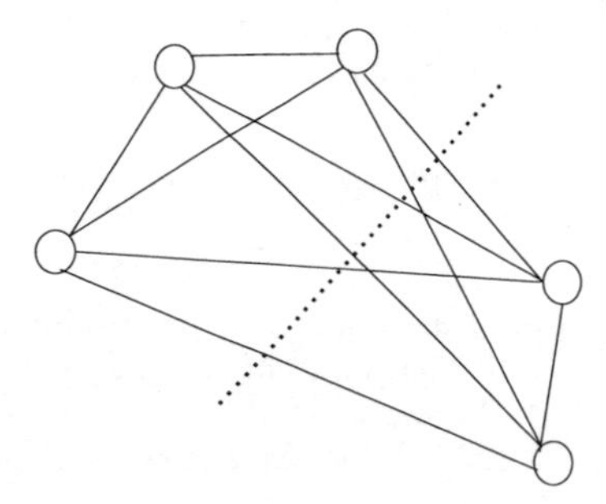

Figure 3.7: The idea behind spectral clustering: Objects to be clustered are symbolized as nodes of the a graph, and their degrees of dissimilarity are depicted by the length of the connecting edges (i.e. longer edges indicate higher dissimilarity). The dashed line shows a good split of the graph that on one hand maximizes the within-cluster and on the other hand minimizes the between-cluster similarity.

3.6.4 Spectral Clustering

Despite the popularity of the k-means clustering algorithm it has the problem that it is prone to local minima, since the optimization of (3.37) is non-convex. Spectral clustering techniques were therefore developed to overcome this problem by relaxing the original problem in a way that global optimal solutions can be found.

The simplest way to introduce spectral clustering is to start with the special case of a partitioning of the data into two classes, i.e. we want to learn a clustering function $f : \mathcal{X} \to \{+1, -1\}$. Let $\mathbf{W} \in \mathbb{R}^{n \times n}$ denote some symmetric similarity matrix with $W_{ii} = 0$. Obviously, a special case is $\mathbf{W} = \mathbf{K} - \text{diag}(\mathbf{K})$, where *diag* denotes an operator that returns the diagonal matrix of its argument. The similarities W_{ij} can be interpreted as weights on the edges connecting nodes i and j in a graph. If $W_{ij} = 0$, then there is no edge from i to j, i.e. $\mathbf{W}$ plays the role of an adjacency matrix. In this way the problem of clustering can be viewed as the problem of finding an optimal cut in a graph (Fig. 3.7). Let $d_i = \sum_j W_{ij}$, called the *degree* of node i, and $\mathbf{D} \in \mathbb{R}^{n \times n}$ a diagonal matrix with $D_{ii} = d_i$. Furthermore, let $\mathbf{y} = (y_1, ..., y_n)^T \in \{+1, -1\}^n$ a vector of labels that are assigned to the data items by the clustering algorithm. Then the *cut-cost* of splitting the graph in two partitions according to a given clustering is given as (c.f. Shawe-Taylor and Cristianini, 2004):

$$\sum_{y_i \neq y_j}^{n} W_{ij} = \frac{1}{2}\left(\sum_{i,j=1}^{n} W_{ij} - \sum_{i,j=1}^{n} y_i y_j W_{ij}\right) = \frac{1}{2}\left(\mathbf{1}^T\mathbf{W}\mathbf{1} - \mathbf{y}^T\mathbf{W}\mathbf{y}\right) = \frac{1}{2}\mathbf{y}^T(\mathbf{D} - \mathbf{W})\mathbf{y} \tag{3.39}$$

where $\mathbf{1} \in \mathbb{R}^n$ denotes a vector of ones. The cut-cost should be normalized in some way by the sizes of the clusters to prefer balanced clusterings. Shi and Malik (2000) proposed the so-called *normalized cut* between c and its complement $\bar{c}$, which they define as

$$\sum_{y_i \neq y_j}^{n} W_{ij} \left(\frac{1}{|c|} + \frac{1}{|\bar{c}|} \right) \tag{3.40}$$

where $|c| = \sum_{y_i = c} d_i$ is the *volume* of cluster c. In principle, finding a vector $\mathbf{y}^*$ minimizing the normalized cut criterion is a combinatorial problem, which is known to be NP complete (Chung, 1997). If, however, $\mathbf{y}$ is allowed to be a real valued vector, then an optimal $\mathbf{y}^*$ can be computed as the solution of the following optimization problem:

$$\min_{\mathbf{y}} \mathbf{y}^T(\mathbf{D} - \mathbf{W})\mathbf{y} \tag{3.41}$$
$$\text{subject to } \mathbf{y}^T\mathbf{D}\mathbf{y} = 1$$

The constraint of the optimization problem not only guarantees the solution $\mathbf{y}^*$ to be nonzero, but also incorporates the normalization, because in the ideal binary case we would have $\mathbf{y}^T\mathbf{D}\mathbf{y} = \sum_{y_i = y_j = 1} D_{ij} + \sum_{y_i = y_j = -1} D_{ij} = |c| + |\bar{c}|$. The optimal solution $\mathbf{y}^*$ can be computed as the eigenvector corresponding to the second smallest eigenvalue of the generalized eigenvector problem $(\mathbf{D} - \mathbf{W})\mathbf{y} = \lambda\mathbf{D}\mathbf{y}$ (Shi and Malik, 2000). A 2-way clustering is then performed by thresholding $\mathbf{y}^*$ appropriately. $\mathbf{D} - \mathbf{W}$ is the so-called *Laplacian* matrix of the graph.

Instead of minimizing the cut cost $\mathbf{y}^T(\mathbf{D} - \mathbf{W})\mathbf{y}$ with appropriate constraints, one can also **maximize** $\mathbf{y}^T(\mathbf{D}^{-1}\mathbf{W})\mathbf{y}$ subject to $\|\mathbf{y}\|^2 = 1$ (Meila and Shi, 2001a):

$$\max_{\mathbf{y}} \mathbf{y}^T(\mathbf{D}^{-1}\mathbf{W})\mathbf{y} \tag{3.42}$$
$$\text{subject to } \mathbf{y}^T\mathbf{y} = 1$$

Thus the optimal $\mathbf{y}^*$ is given by the eigenvector corresponding to the largest eigenvalue of the eigenvector problem $(\mathbf{D}^{-1}\mathbf{W})\mathbf{y} = \lambda\mathbf{y}$. Meila and Shi (2001b) interpret $\mathbf{D}^{-1}\mathbf{W}$ as the transition probability matrix of a one-step random walk on the graph and show that the optimization problems (3.41) and (3.42) lead to the same eigenvectors.

Additionally to the normalized cut criterion other objective functions were proposed as well: E.g. Ding et al. (2001) use the so-called *min-max-cut*

$$\sum_{y_i \neq y_j}^{n} W_{ij} \left(\frac{1}{\sum_{y_i = y_j = 1} W_{ij}} + \frac{1}{\sum_{y_i = y_j = -1} W_{ij}} \right) \tag{3.43}$$

to obtain a better balanced clustering and arrive at the eigenvector problem $\mathbf{D}^{-1/2}\mathbf{W}\mathbf{D}^{-1/2}$ $\mathbf{y} = \lambda\mathbf{y}$. The matrix $\mathbf{D}^{-1/2}\mathbf{W}\mathbf{D}^{-1/2}$ is closely related to the *normalized Laplacian* $\mathbf{I} - \mathbf{D}^{-1/2}\mathbf{W}\,\mathbf{D}^{-1/2}$ of a graph.

For the problem of finding $k > 2$ clusters two basic principles have been proposed:

- Recursively use a 2-way clustering to split clusters (Shi and Malik, 2000; Kannan and Vempala, 2000). Thereby the problem arises, which cluster to split next (c.f. Ding et al., 2001).
- Compute K eigenvectors and perform e.g. k-means clustering on the coordinates of the eigenvectors (Meila and Shi, 2001b; Ng et al., 2002). The motivation for this is that the coordinates of the eigenvectors show the degree of membership of a point to the K clusters. By performing k-means clustering on the coordinates points are grouped together, which have a similar membership profile to the clusters.

In this thesis we utilized the spectral clustering algorithm of Ng et al. (2002), which applies the second principle and works as follows:

1. Compute the K eigenvectors $\mathbf{u}_1, ..., \mathbf{u}_K$ corresponding to the K largest eigenvalues of $\mathbf{D}^{-1/2}\mathbf{W}\mathbf{D}^{-1/2}$.
2. Form the matrix $\mathbf{U} = [\mathbf{u}_1\mathbf{u}_2...\mathbf{u}_K]$ by stacking the eigenvectors into columns and normalize each row of $\mathbf{U}$ to unit length.
3. Perform k-means clustering on the rows of $\mathbf{U}$ and assign each point $\phi(x_i)$ to its corresponding cluster based on the result.

If the clustering was successful, the kernel matrix should reveal a block structure after re-ordering the rows and columns with respect to the cluster membership of the data objects (Fig. 3.8). This indicates that similar data items are grouped together, whereas dissimilar ones are put into separate clusters.

3.7 Summary

We presented those kernel-based learning algorithms, which play a crucial role throughout this thesis:

- SVMs for classification rely on the idea of a maximum margin hyperplane classification in feature space. The optimal hyperplane is uniquely determined and can be found by quadratic programming optimization techniques. The solution is sparse and only relies on those examples that are support vectors. Nonlinear classification and classification of arbitrary objects is achieved by introducing kernels. A soft margin SVM allows for training errors. There exist two basic formulations for soft margin SVMs: C-SVMs and ν-SVMs, which only differ in the way they parameterize the trade-off between margin maximization and training error minimization. The parameter ν can be understood as a lower bound for the fraction of examples, that are support vectors and an upper bound on the fraction of examples that are

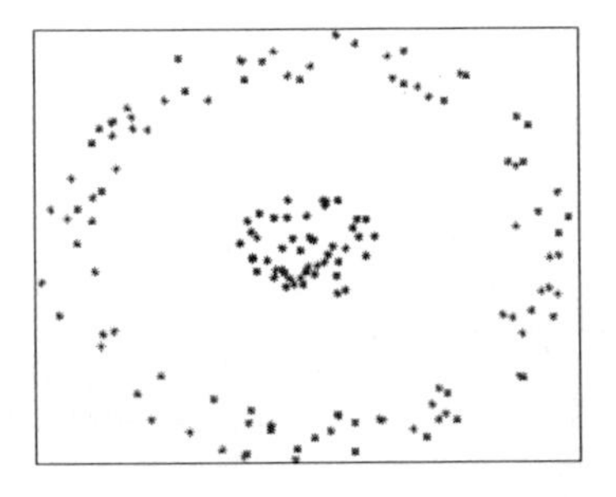

(a) Two nested rings.

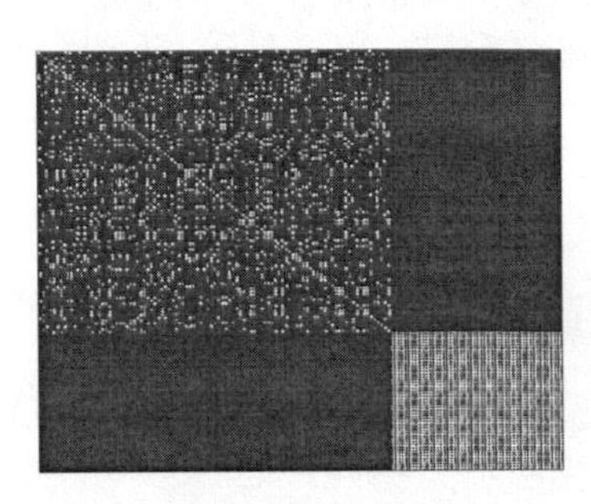

(b) Corresponding kernel matrix after clustering: lighter pixels indicate higher similarity.

Figure 3.8: An example clustering of two rings: The kernel matrix shows a clear block structure.

margin errors. The simplest way to perform multi-class classification for SVMs is to employ a one-versus-rest or a one-against-one strategy. The generalization performance of an SVM can be estimated efficiently from training data only by means of the span-rule or by means of the radius-margin bound.

- Kernel Ridge Regression is a straight forward application of the Tikhonov regularization framework to perform linear regression in feature space. The solution can be found by a simple matrix inversion, but is not sparse.

- SVMs for regression, also called SVR, were developed to introduce the benefit of sparsity into kernel-based regression estimation. In contrast to Kernel Ridge Regression, SVR utilizes a special loss function, the ϵ-insensitive loss, to achieve this purpose. Regularized risk minimization leads to a quadratic programming optimization problem. The solution is uniquely determined and only relies on the support vectors, which are those examples that lie outside or exactly on the ϵ-tube around the regression function. The generalization performance of a SVR can be estimated efficiently with a bound similar to the radius-margin for classification SVMs.

- KCCA is a method to extract correlated features from data, which is represented in two different feature spaces. Hence, the basic assumption is that indeed both representations have a certain common structure. The problem is solved by constructing orthogonal coordinate systems in both feature spaces, such that the new axis successively extract the directions with the highest correlation between both views. This can be achieved by solving a generalized eigenvector problem. A dual formulation allows for the usage of kernels.

- Clustering in feature spaces can be performed by classical methods, like the linkage algorithms or, slightly more complicated, by k-means. This leads to the dual k-means algorithm. Spectral clustering methods are an alternative, which are not as greedy as k-means. The idea is to construct an optimal cut in a graph. In case of a 2-way clustering the solution can be found by thresholding the solution of an eigenvector problem appropriately. As the single coordinates of the eigenvectors indicate the degree of membership of a point to the clusters, with more clusters a common strategy is to perform k-means clustering on the coordinates of the eigenvectors. The final clustering is then computed based on this result.

A crucial question, which we did not tackle within this Chapter, is how parameters for the individual algorithms can be chosen in an optimal way. This leads to the problem of model selection. Concerning SVMs and SVR, in the next Chapter we try to find a fairly general solution for this problem.

Chapter 4

Model Selection for Support Vector Machines

4.1 Introduction

A special strength of SVMs is the use of a kernel function to introduce nonlinearity and to deal with arbitrarily structured data. Usually the kernel depends on certain parameters, which, together with other parameters of the SVM, have to be tuned to achieve good results. For instance, in classification using a simple RBF kernel we have the width σ of the RBF function and the soft margin parameter C to be tuned, while in regression we have also the additional parameter ϵ to control the width of the ϵ-tube. If we consider arbitrary kernel functions, like for strings, trees or graphs, the number of parameters D and hence the size of the parameter space $\mathcal{P} = \mathbb{R}^D$ may also be higher. The importance of the tuning procedure is an often neglected issue. Supposed we are given some measure of quality $Q : \mathcal{P} \to \mathbb{R}$, which for each set of parameters $\mathbf{p}$ estimates the generalization error $Q(\mathbf{p})$ of the SVM, then we are interested in those parameters $\hat{\mathbf{p}}$, for which Q becomes minimal, i.e.

$$\hat{\mathbf{p}} = \arg\min_{\mathbf{p}} Q(\mathbf{p}) \tag{4.1}$$

This is the so-called *model selection* problem, and ideally we would like to find the global optimum of (4.1). Note that we cannot assume the different parameters to be independent. This prevents us from tuning each parameter separately from the rest.

The standard method to deal with problem (4.1) is to use a simple grid search on the log-scale of the parameters in combination with cross-validation on each candidate vector $\mathbf{p}$ of parameters. If, however, the number of parameters becomes higher, this leads to an explosion of necessary SVM trainings. Consider e.g. we test $\log_2 C \in \{-2, ..., 14\}, \log_2 \epsilon \in \{-8, ..., -1\}, \log_2 \sigma \in \{-8, ..., 8\}$ in a regression experiment with a simple RBF kernel. This would lead to $17 \cdot 8 \cdot 17 = 2312$ cross-validation runs. Of course there are ways to speed things up a little bit, since we can roughly estimate C from the range of function values, the order of σ from the median (or mean) distance between points in the original space and ϵ from the noise level in our data (Cherkassky and Ma, 2004). However, even then the general problem remains, because all these estimates can just be viewed as first guesses. The situation becomes much more complicated, if we consider kernels on non-vectorial data.

For SVMs in classification, Chapelle et al. proposed a very efficient approach for model selection by performing a gradient descent on either the radius-margin or the span-bound (Chapelle et al., 2002; Chapelle and Vapnik, 2000; c.f. Chapter 3.2.2). A drawback of this method is, however, the need for a gradient computation, which for general kernel functions might either not be possible or at least be very difficult (think e.g. about kernels for graphs — c.f. Chapter 6). Additionally, the radius-margin and the span-bound are just upper bounds on the true risk and the leave-one-out error, respectively, and, despite good experimental results, in general it is unknown how close these bounds are. It is also worth mentioning that the gradient descent may get stuck in local optima. In case of Support Vector Regression (SVR) the method is not applicable, and for the multi-class case the usage is not totally clear.

Aside from this approach, there exist several methods for tuning the SVM/SVR parameters C (and ϵ), if the kernel parameters are given. In Kwok (2000) and Law and Kwok (2001) the authors use a Bayesian interpretation of the SVM/SVR to estimate C (and ϵ) via MacKay's evidence procedure (MacKay, 1992). In a recent publication Hastie et al. (2004) introduce an algorithm that can fit the entire path of SVM solutions in classification for every value of the parameter C with essentially the same computational cost as fitting one SVM model.

In this Chapter, we want to consider the general case of tuning parameters for a kernel function, which depends on several parameters (and is not necessarily differentiable) and additional SVM parameters in regression as well as classification. Our goal is to have a general approach for the model selection problem. To our best knowledge the only work covering this situation is the paper by Momma and Bennett (2002). Their proposed algorithm employs the pattern search method by Dennis and Torczon (1994), a derivative-free method, which, beginning from a starting point, investigates the neighbors of a parameter vector. Thereby the transition from one point in search space to another is defined by a fixed neighbor sampling pattern and the length of the search step. The length of the search step is reduced at each iteration until convergence is reached. The pattern search is started at a random location. To avoid local optima and in order to increase robustness of the method Momma and Bennett propose to use bagging or model averaging.

In contrast, our idea is to treat the model selection problem directly as a global optimization problem (c.f. Fröhlich and Zell, 2005). The general intuition is to learn a model, namely an Online Gaussian Process (GP — Csato and Opper, 2002), from the points in parameter space we have already visited. We will argue, that in contrast to the original SVM model, training and testing of the Online GP can be performed very cheaply. New points in parameter space are sampled according to the expected improvement criterion as defined by Jones et al. in the EGO algorithm (Jones et al., 1998), which balances global and local search.

In the following Section we will describe our method in detail. Section 4.3 contains extensive experimental evaluations in comparison to a grid search and the pattern search approach with a following discussion in Section 4.4. Finally, in Section 4.5 we conclude.

4.2 The Algorithm

4.2.1 Our Method

Theoretical Background: Online Gaussian Processes Online Gaussian Processes have been introduced recently as an elegant extension of Gaussian Processes (GPs — e.g. MacKay, 1997; Willams, 1997) to the scenario of online learning regression and classification functions (Csato and Opper, 2002).

A Gaussian Process is a stochastic process, which in our case is defined over the function values of an unknown regression function $f : \mathcal{X} \rightarrow \mathbb{R}$. A stochastic process in general is a collection of random variables $\{G(y)|y \in \mathcal{Y}\}$ where $\mathcal{Y}$ is some domain (in our case $\mathcal{Y} = \mathbb{R}$). The stochastic process is defined by giving the probability distribution for every finite subset of variables $G(y_1), ..., G(y_n)$ in a consistent manner. A Gaussian Process can be fully specified by its mean function $\mu_f(\cdot) = \mathbf{E}[G(\cdot)]$, i.e. the expectation over the function values, and its covariance or kernel function $k : \mathcal{X} \times \mathcal{X} \rightarrow \mathbb{R},\ k(x, x') = \mathbf{E}[(G(f(x)) - \mu_f(f(x)))(G(f(x')) - \mu_f(f(x')))]$, which describes the degree of correlation between function values $f(x)$ and $f(x')$ depending on their locations x and x'. Any finite set of function values will have a joint Gaussian distribution. The standard setting is $\mu_f \equiv 0$, which implies that we have no prior knowledge about particular estimations, but generally prefer small values. In practice that means one should normalize the target values in the training data to mean 0 and remove any known trend in the data.

In case of batch learning we are given a fixed dataset $\mathcal{D} = \{(x_i, y_i)|x_i \in \mathcal{X}, y_i \in \mathbb{R}, i = 1, ..., n\}$ of n data points and a fixed covariance function $k : \mathcal{X} \times \mathcal{X} \rightarrow \mathbb{R}$, which maps pairs of x-values to the covariances of their corresponding function values. In a Bayesian inference framework (e.g. Bishop, 1995) the form of the covariance function can be interpreted as a prior over possible regression functions we can learn. An often used covariance function is e.g. the Gaussian kernel. In this case it is a standard result that the posterior predictive distribution $p_{post}(y^*|\mathcal{D})$ corresponding to some test point x^* is $y^* \sim \mathcal{N}(\hat{\mu}(x^*), \hat{\sigma}(x^*))$, and $\hat{\mu}(x^*)$ and $\hat{\sigma}(x^*)$ can be computed by simple matrix manipulations from the kernel matrix and the observed output values y_i (MacKay, 1997; Willams, 1997).

In case of online learning we are given data $\mathcal{D} = \{(x_i, y_i)|x_i \in \mathcal{X}, y_i \in \mathbb{R}, i = 1, ..., t\}$ up to the current time step t. We want to update our GP model which has been constructed on $\mathcal{D}$ before, as soon as the next example $z_{t+1} = (x_{t+1}, y_{t+1})$ arrives. Let $\hat{p}_t$ denote the Gaussian Process approximation after processing t examples and $\mathbf{y} = (y_1, ..., y_t)^T$. We use Bayes' rule to derive the updated posterior distribution over predicted values y^* at some point x^* (Csato and Opper, 2002):

$$p_{post}(y^*|z_{t+1}) = \frac{p(z_{t+1}|y^*)\hat{p}_t(y^*)}{\int p(z_{t+1}|\mathbf{y})\hat{p}_t(\mathbf{y})d\mathbf{y}} \tag{4.2}$$

Usually the direct computation of (4.2) will be intractable, especially because p_{post} is no longer Gaussian. However, Csato and Opper show that indeed the **expected** model value $\hat{\mu}_t(x^*)$ corresponding to the Gaussian Online Process at time t can be written as

$$\hat{\mu}_t(x^*) = \sum_{i=1}^{t} k(x^*, x_i)\alpha_t(i) \tag{4.3}$$

with coefficients $\alpha_t(i)$, which can be computed via a recursive update formula using the covariances $\mathbf{k}_t = (k(x^*, x_1), ..., k(x^*, x_{t1}))^T$, i.e. the model can be updated as soon as a new example arrives by using the covariances of the new example with the old ones. A special feature of Gaussian Processes is the fact, that besides the expected model output we can receive an estimation of the variance $\hat{\sigma}_t^2(x^*)$ of the model at point x^*. This can be computed from the covariances $\mathbf{k}_t$ and $k(x^*, x^*)$:

$$\hat{\sigma}_t^2(x^*) = k(x^*, x^*) + \mathbf{k}_t^T \mathbf{C}_t \mathbf{k}_t \tag{4.4}$$

Here $\mathbf{C}_t$ is a matrix of coefficients not depending on x, x^*, which can be updated in a recursive fashion as well.

Application to SVM Model Selection Our goal is to learn an Online GP model from the points in parameter space we have already visited. Beginning from a number of initial points that can be determined by Latin hypercube sampling, we update our model at each search step. That means at each search step we refine our Online GP regression model $f : \mathcal{P} \to \mathbb{R}$ of the error surface of the SVM model. Following Jones et al. (1998) we set the number of initial points to be around $10D$. We choose a Gaussian covariance function for the Online GP with hyperparameters being adapted by maximum likelihood (c.f. MacKay, 1997).

At a first glance one might ask what we win by modeling the error surface of a SVR model in parameter space via another regression model. Indeed, there is only a gain if we assume that the training time for the Online GP is very small compared to that of the full SVM model. However, this is a reasonable assumption, since the number of training points for the Online GP and hence the number of evaluations of the error surface of the SVM model mainly depends on the dimensionality of the parameter space, which is very small compared to the number of training points for the SVM.

There is the remaining question how, given our current Online GP model, we can find the next sample point in parameter space. Here we use the *expected improvement*

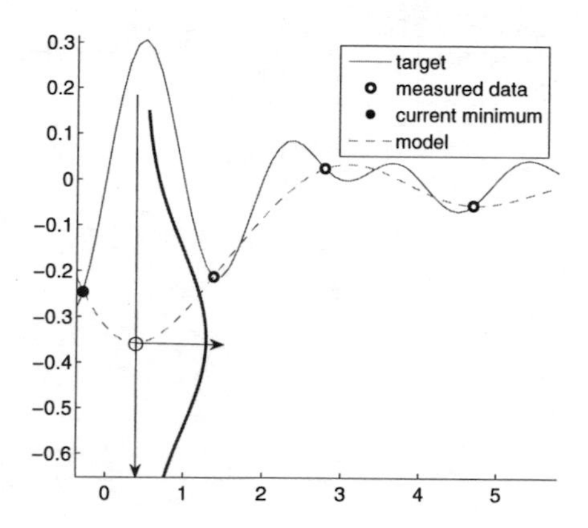

Figure 4.1: The uncertainty of the model at some point $\mathbf{p}$ (white circle) can be modeled as the realization of a random variable.

criterion as defined by Jones et al. in the EGO algorithm (Jones et al., 1998): Formally, the improvement $I(\mathbf{p})$ at some point $\mathbf{p}$ in parameter space is defined as $I(\mathbf{p}) = \max(0, Q_{min} - Y)$, where Q_{min} is our current minimum of the estimated generalization error of the SVM model and $Y = \mathcal{N}(\hat{\mu}(\mathbf{p}), \hat{\sigma}(\mathbf{p}))$ (see Fig. 4.1). Note, that $\hat{\mu}(\mathbf{p})$ is the expected model value at $\mathbf{p}$. Indeed, I is a random variable, because Y is a random variable (it models our uncertainty about $\hat{\mu}(\mathbf{p})$). To obtain the expected improvement we simply take the expectation

$$\mathbf{E}[I(\mathbf{p})] = (Q_{min} - \hat{\mu}(\mathbf{p}))\Phi\left(\frac{Q_{min} - \hat{\mu}(\mathbf{p})}{\hat{\sigma}(\mathbf{p})}\right) + \hat{\sigma}(\mathbf{p})\phi\left(\frac{Q_{min} - \hat{\mu}(\mathbf{p})}{\hat{\sigma}(\mathbf{p})}\right) \tag{4.5}$$

where Φ and ϕ are the standard normal distribution and density function.

The expected improvement can be viewed as a function $\mathbf{E}[I(\mathbf{p})] : \mathcal{P} \rightarrow \mathbb{R}$ and can be evaluated cheaply over the whole parameter space in contrast to the costly evaluations of the error surface of the SVM model (see Fig. 4.2). This is, because we just need the predictions of the already trained Online GP, which is fast. Now the idea is to sample at that point next, for which the expected improvement becomes maximal, i.e. we are looking for

$$\tilde{\mathbf{p}} = \arg\max_{\mathbf{p}} \mathbf{E}[I(\mathbf{p})] \tag{4.6}$$

This is a global optimization problem for itself, but in contrast to our original problem (4.1) the solution can be computed cheaply, e.g. by the DIRECT algorithm (Jones et al.,

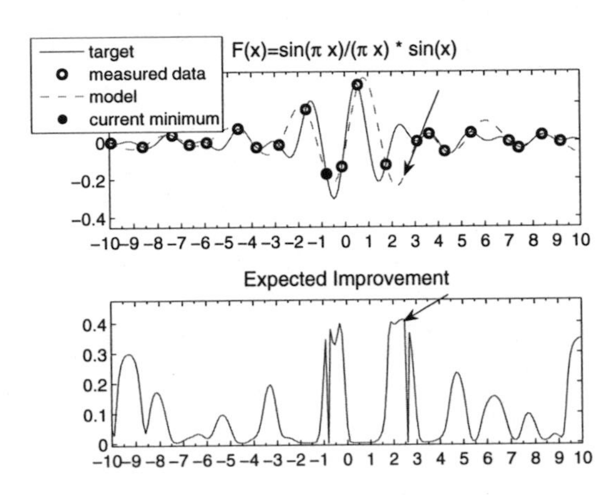

Figure 4.2: The expected improvement for an example function. The next sample point would be set where the expected improvement is highest (arrows).

1993), because of the reasons described above. DIRECT is a sampling algorithm which requires no knowledge of the objective function gradient. Instead, the algorithm samples points in the domain, and uses the information it has obtained to decide where to search next. The DIRECT algorithm will globally converge to the maximal value of the objective function (Jones et al., 1993). The name DIRECT comes from the shortening of the phrase "DIviding RECTangles", which describes the way the algorithm moves toward the optimum.

Now we have all ingredients, which describe our algorithm to tune SVM and kernel parameters. In pseudo-code 1 we give an overview over the whole procedure, which we call **EPSGO** (**E**fficient **P**arameter **S**election via **G**lobal **O**ptimization): Beginning from an initial Latin hypercube sampling we train an Online GP, look for the point with the maximal expected improvement, sample there and update our Online GP. Thereby it is not so important that our Online GP really correctly models the error surface of the SVM in parameter space, but that it can give us information about potentially interesting points in parameter space where we should sample next. We continue with sampling points until some convergence criterion is met. In our case we decided to stop, if the difference between the current best value of Q and the old one is small, and the squared difference between the maximal expected improvement and the average expected improvement is less than 10% of the standard deviation of the expected improvement. These statistics of the expected improvement can be computed using Latin hypercube sampling over the

Algorithm 1 The EPSGO algorithm

```
Input: function Q to measure gen. error
       l,u: parameter bounds
Output: Q_min, p̂, # Q-evaluations (n_eval)

D = dim(P)
create N = 10D sample points p_1,...,p_N
   in [l,u] using Latin hypercube sampl.
compute Q(p_i), i = 1,...,N
Q_min = min_i Q(p_i)
p̂ = arg min_i Q(p_i)
train Online GP
n_eval = N
REPEAT
  p̃ = arg max_p E[I(p)] (computed by DIRECT)
  compute std. dev. and mean of E[I(p)]
  Q_new = Q(p̃)
  if Q_new < Q_min
    Q_min = Q_new
    p̂ = p̃
  end
  update Online GP
  n_eval = n_eval + 1
UNTIL convergence
```

whole parameter space (– note again, that these calculations are very fast). The idea is to make sure, that the expected improvement over the whole space is almost equally small. To prevent long searches we also stopped, if Q_{min} did not change during the last 10 iterations.

4.2.2 Relationship to Existing Global Optimization Methods

The EPSGO algorithm can be viewed as a special variant of the EGO algorithm proposed by Jones et al. (1998) for globally optimizing costly target functions. While the EGO algorithm uses a DACE model, which has to be retrained from time to time, we use an Online GP which can be updated efficiently in $\mathcal{O}(N)$ time (N = number of parameter vector tested until time step t) as soon as a new sample point arrives. Jones et al. use a branch-and-bound algorithm to maximize the expected improvement, while we take the DIRECT algorithm, which has no need for the computation of bounds.

Table 4.1: Computation times for optimizing difficult test functions with DIRECT. The precision (absolute deviance from the global optimum) is reported in the last column.

Function	dim.	time (s)	#func. eval.	precision
Goldstein-Price	2	0.05	191	$< 10^{-4}$
Branin	2	0.13	971	$< 10^{-6}$
Michaelwicz	5	8.15	13.961	$< 10^{-10}$

4.2.3 Computational Analysis

Computationally, clearly the most critical part in the EPSGO algorithm is the evaluation of the quality function Q (e.g. the cross-validation error) itself, which is dependent on the dataset, the kernel function and the employed SVM implementation[1]. Jones et al. (1998) do not prove any convergence properties or computational complexities of EGO. Therefore, in the experimental part we resort to measuring the number of Q-function evaluations needed until our termination criterion is met, which is in accordance to Jones et al. (1993, 1998). To give a better impression on the performance of the DIRECT algorithm, we conducted experiments with three well known test functions from global optimization theory: the Goldstein and Price function, the Branin function and the Michaelwicz function (see Appendix for details). In Table 4.1 we show the number of function evaluations and the computation time for optimizing these difficult test functions. Obviously, the computation time is rather small to achieve a very high accuracy of optimization. In conclusion it is fair to say that all computations within the EPSGO algorithm are extremely cheap except for the the evaluation of the quality function Q.

4.3 Experiments

4.3.1 Datasets and Experimental Setup

We evaluated our method on 5 classification and 4 regression datasets from the UCI machine learning repository, which is a well known collection of public available benchmark datasets[2]: the Iris, Glass, Wine, Cleveland Heart and Wisconsin Breast Cancer dataset for classification, and the Boston Housing, Triazines, Pyrimidines and Auto-MPG dataset for regression. These are often used benchmark problems for machine learning algorithms and are therefore taken in this work as well. Information on the individual datasets can be found in Appendix A of this thesis. In this context it should be emphasized again that our algorithm is intended to be of general usage and therefore not restricted to chemo- or

[1]The cost for one SVM training is usually $\mathcal{O}(dn^3)$ with n being the number of training points and d the number of features

[2]http://www.ics.uci.edu/~mlearn/MLSummary.html

bioinformatics applications.

Additionally to the above described datasets, we included a noisy sinc ($\sin(x)/x$) function (10% normally distributed random noise) in the range -15 to +15 as a standard example for regression and a variant of Wieland's two spirals with 200 points per class and 3 coils (Fig. 4.3.2). Both are very often used toy problems in the machine learning literature.

All data were normalized to mean 0 and standard deviation 1 before training. On all experiments we took a RBF kernel with width σ. Additional SVM parameters were C for classification, and C and ϵ for regression tasks. An one-against-one approach was used to deal with the multi-class case in the classification experiments. As quality measure Q for each set of parameters we used 5-fold cross-validation. We compared our EPSGO method to a grid search with $\log_2 C \in \{-2, ..., 14\}, \log_2 \sigma \in \{-5, ..., 7\}$ and (for the regression datasets) $\log_2 \epsilon \in \{-8, ..., -1\}$ by means of an extra level of 5-fold cross-validation. On the Triazines and Pyrimidines dataset we used the predefined training and testing folds. We also compared our algorithm to the pattern search method of Momma and Bennett with and without bagging (PS/PS-Bag) and a bagged version of EPSGO (EPSGO-Bag). The bagging was performed over 5 models, i.e. we have 5 times the computational effort as without bagging.

4.3.2 Results

Figures 4.4, 4.5 and Table C.1 in the Appendix show the results of the comparisons of EPSGO to a grid search and the pattern search method without bagging (PS). On the classification data EPSGO needed less than 10% of the search steps of a grid search and lead to almost identical results. On the regression data the advantage became even more obvious. Here we needed around 1% of the search steps, and in 1 case (noisy sinc function) we obtained a significantly better result than with grid search. Thereby significance was tested by means of a two-tailed paired t-test at 5% significance level. Compared to the PS method, EPSGO in 7 cases obtained significantly lower error rates, which shows that the pattern search method without bagging often suffers from the problem of local minima and lacks statistical robustness. At the same time on average the number of search steps performed by EPSGO was comparable to those performed by the PS method (Table 4.2).

The next comparisons were made between the bagged version of the pattern search method (PS-Bag), the EPSGO algorithm and a bagged version of the EPSGO algorithm (EPSGO-Bag — Fig. 4.6 and 4.7, Table C.2 in the Appendix). Obviously, PS-Bag needs 5 times the number of search steps of PS. However, this significantly higher amount of computation time only in 4 cases leads to a significant lower error rate of PS-Bag compared to EPSGO. In 6 cases EPSGO was significantly better. The bagged version of EPSGO, EPSGO-Bag, in 6 cases obtained significant lower error rates, while never being significantly worse than PS-Bag. This again shows the higher robustness and insensitivity to local minima of the EPSGO approach compared to the PS approach. The number of

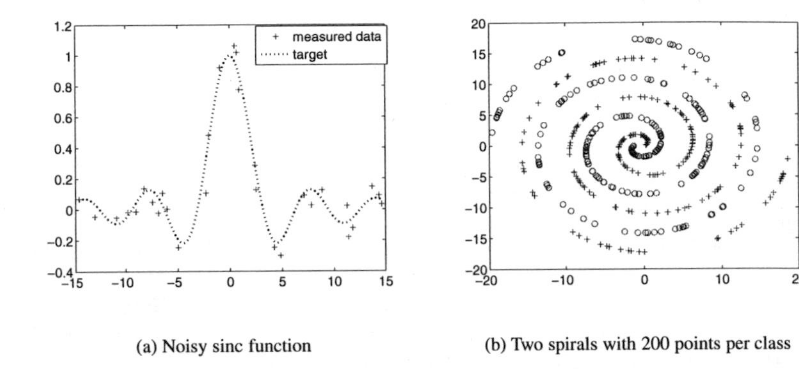

(a) Noisy sinc function

(b) Two spirals with 200 points per class

Figure 4.3: Two toy problems.

Table 4.2: Average number of search steps ($\pm$ standard deviation) needed by the different methods.

Method	Classification	Regression
EPSGO	24 ± 4	33 ± 2
PS	22 ± 0	32 ± 0
EPSGO-Bag	120 ± 23	206 ± 33
PS-Bag	110 ± 0	160 ± 0
grid search	384 ± 0	3072 ± 0

search steps performed by EPSGO-Bag on the regression data on average was slightly higher than those performed by PS-Bag (Table 4.2). Compared to EPSGO, EPSGO-Bag obtained significantly lower error rates in 3 cases.

4.4 Discussion

All in all we see that EPSGO/EPSGO-Bag achieved a substantially higher robustness and insensitivity to local minima compared to the PS/PS-Bag. The obvious reason for this is, that the pattern search depends on the random initialization of just 1 starting point. This dependency can be reduced by using bagging, as proposed by Momma and Bennett, which imposes a much higher computational burden. However, it is still a clearly observable disadvantage compared to our EPSGO method. It is remarkable, that EPSGO without using bagging (and hence being much faster than PS-Bag) in our experiments significantly

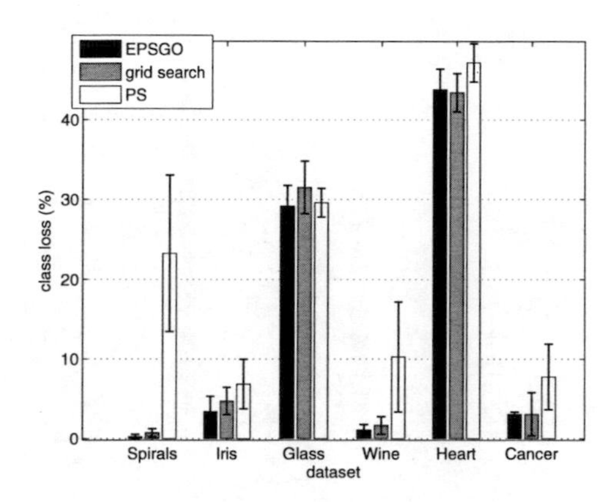

Figure 4.4: 5-fold cross-validation errors for the classification problems (class loss in %). Comparison of EPSGO vs. grid search and pattern search (PS).

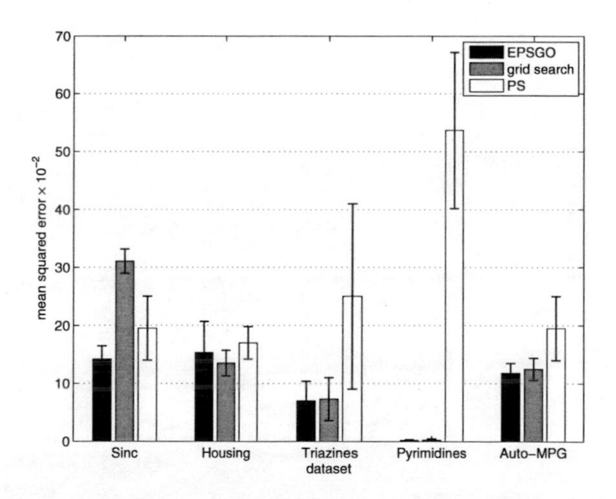

Figure 4.5: 5-fold cross-validation errors for the regression problems (mean squared error). Comparison of EPSGO vs. grid search and pattern search (PS). For better display for the Triazines dataset the MSE is scaled by 100.

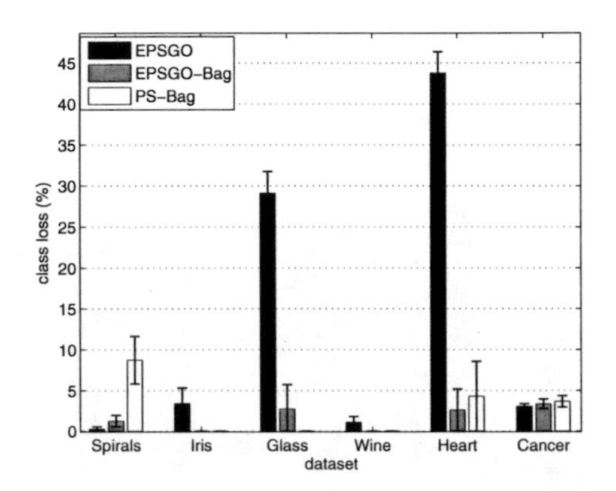

Figure 4.6: 5-fold cross-validation errors for the classification problems (class loss in %). Comparison of EPSGO-Bag vs. EPSGO and bagged pattern search (PS-Bag).

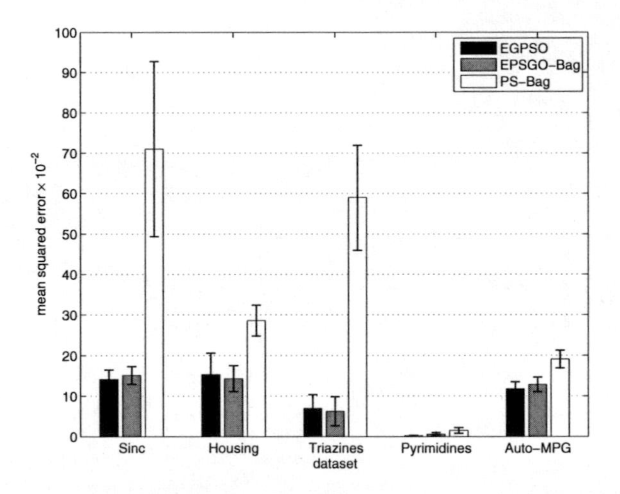

Figure 4.7: 5-fold cross-validation errors for the regression problems (mean squared error). Comparison of EPSGO-Bag vs. EPSGO and bagged pattern search (PS-Bag). For better display for the Triazines dataset the MSE is scaled by 100.

outperformed PS-Bag more often than vice versa.

For practical purposes it depends on the user's point of view, whether it is worth taking the higher computational effort of bagging compared to the usual win of lower error rates. If one wants to use bagging, then the EPSGO-Bag method offers a stable and robust way, which at the same time is not very sensitive to local minima, to obtain low error rates while still being very efficient compared to a grid search. Compared to the bagged version of the pattern search EPSGO-Bag on average needs a slightly higher number of search steps. If one does not want to spend the extra time for bagging, then EPSGO without bagging is a good alternative. It leads to results comparable to a grid search, while performing a much faster search in parameter space. This advantage increases, the more parameters we have to tune.

4.5 Conclusion

We proposed an efficient method to perform model selection for SVMs, which is not dependent on special properties of the kernel, e.g. differentiability. Our method is very general and applicable for classification as well as regression tasks. It is a special variant of the EGO algorithm by Jones et al. (1998) used in global optimization. We use an Online Gaussian Process to learn a model in parameter space and sample new points, where the expected improvement is maximal. To efficiently find the maximum of the expected improvement we employ the DIRECT algorithm. Comparisons of our method to a usual grid search showed the high win of performance with regard to the number of search steps needed. At the same time we obtained error rates, which were at least as good. Compared to the pattern search method of Momma and Bennett our approach revealed a better statistical robustness and insensitivity to local minima. This was observable even, if bagging was used in combination with the pattern search.

Chapter 5

Descriptor Selection for Molecules with SVMs

5.1 Introduction

5.1.1 QSPR Models in Virtual Screening

The development of a new drug is a very costly and long lasting process. From the late 19th century until the seventies of the 20th century the strategies to discover new drugs did not change much: New compounds were synthesized and tested on animals or organ preparations following some chemical or biological hypothesis (c.f. Kubinyi, 2004). Frequent animal experiments to test new compounds not only raised ethical questions, but also took a lot more time and money than needed for chemical synthesis. About 30 years ago rational approaches developed with the aim to significantly reduce the amount of compounds to be tested *in vivo*. *In vitro* test systems such as enzyme inhibition or the displacement of radio-labeled ligands in membrane preparations enabled much faster investigation of potential drug candidates.

With the invention of combinatorial chemistry systems 10 to 15 years ago the amount of compounds to be tested has dramatically increased and hence new paradigms in drug discovery were needed. Today compound libraries contain millions of structures, of which, of course, only a very small fraction has the potential for a further development to a new drug. The development of a new drug is hence often compared with finding a needle in a haystack (Kubinyi, 2004). One important strategy to deal with this problem are automated *high-throughput screening* systems (HTS), which today allow the biological screening of several million compounds per 24 hours.

Complementary, there is an increasing interest in so-called *virtual screening* methods (c.f. Böhm and Schneider, 2000) to reduce the size of the haystack a-priori and thus to avoid as many costly and time consuming experiments later on as possible. The goal is to filter out a significant amount of "uninteresting" chemicals, that cannot be used as potential drugs, *in silico* in an early stage of the drug discovery process. Thereby, especially the so-called *ADME* (*A*bsorption, *D*istribution, *M*etabolism, *E*xcretion) properties of a compound are of great interest (Kubinyi, 2002, 2003, 2004; van de Waterbeemd and Gifford, 2003): As most drugs are given orally for reasons of convenience, the compound is dissolved in the gastro-intestinal tract. It then has to be absorbed through the gut wall and pass the liver to get into the blood circulation. The percentage of the compound dose reaching the circulation is called the *bioavailability*. From there, the potential drug will have to be distributed to various tissues and organs in the body. The extend of distribution will depend on the structural and physico-chemical properties of the compound. For some drugs it will be further necessary to enter the central nervous system by crossing the blood-brain barrier. Finally, the chemical has to bind to its molecular target, for example, a receptor or ion channel, and exert its desired action.

The body will eventually try to eliminate a drug. Hence, for many drugs this requires metabolism or *biotransformation*. This takes place partly in the gut wall during absorption, but primarily in the liver. Traditionally, a distinction is made between phase I and phase II metabolism, although these do not necessarily occur sequentially. In phase I metabolism, a molecule is functionalized, for example, through oxidation, reduction or hydrolysis. In phase II metabolism, the functionalized compound is further transformed in so-called conjugation reactions, e.g. glucuronidation, sulfation or conjugation with glutathione.

The clearance of a drug from the body mainly takes place via the liver (hepatic clearance or metabolism, and biliary excretion) and the kidney (renal excretion). The *half-life* ($t_{1/2}$) of a compound is the time taken for its concentration in the blood plasma to be reduced by 50%. It is a function of the clearance and volume of distribution, and determines how often a drug needs to be administered.

QSPR (*Quantitative Structure Property Relationship*) methods try to predict *in silico* various ADME, but also physico-chemical properties, which have an important impact on a drug's pharmacokinetic and metabolic fate in the body. Among others, today models for forecasting oral absorption, bioavailability, degree of blood-brain barrier penetration, clearance and volume of distribution are available. Additionally, there are methods for predicting physico-chemical properties, such as e.g. lipophilicity and water solubility (van de Waterbeemd and Gifford, 2003). Similarly, *QSAR* (*Quantitative Structure Activity Relationship*) methods are used to forecast the biological activity/inactivity of an untested ligand for a target protein. (Kubinyi, 2002, 2003, 2004).

The basic assumption behind all QSAR/QSPR approaches is that the molecular properties in question can be derived from certain aspects of the molecular structure only. This implies that structurally similar compounds have similar biological or physico-chemical properties as well. In practice this supposition is often fulfilled, but there are also counter examples (Kubinyi, 2002, 2003).

Typically, QSAR/QSPR models are derived via machine learning methods. Hence, one needs an abstract representation of a chemical compound in the computer. Classically, this is done by a large amount of *descriptors* (= features in machine learning language), which represent global molecular properties, like the polar surface area (van de Waterbeemd and Gifford, 2003), the distribution of certain physico-chemical properties, like the Radial Distribution Function (RDF) descriptor, the frequency of the occurrence of certain atomic patterns (fingerprints), invariances or characteristics of the molecular graph (topological indices) or others (c.f. Todeschini and Consonni, 2000). In conclusion, for each chemical compound one can calculate hundreds or even thousands of descriptors, which are of potential interest. The bottom line is that each molecule, which by itself is a complex three dimensional and dynamic object, is described in a simplified manner by a vector representation, which allows the easy use of any classical data mining method. Often employed statistical methods for learning a QSAR/QSPR model are multiple linear regression, partial least squares regression (PLS), neural networks or, more recently, SVMs (van de Waterbeemd and Gifford, 2003; Kubinyi, 2003; Byvatov et al., 2003).

Besides classical descriptor-based QSAR/QSPR methods, there are also so-called *3D QSAR* approaches, such as *comparative molecular field analysis* (CoMFA — Cramer et al., 1988), where molecular properties are derived from fields calculated on a three dimensional grid, on which chemically related compounds are aligned. 3D QSAR methods usually come with significantly higher computational costs than classical QSAR/QSPR approaches and are not further considered in the context of this thesis.

5.1.2 The Descriptor Selection Problem

If we restrict ourselves to a classical descriptor-based representation of molecules, then a crucial question is what descriptors one should use to obtain a "good" QSAR/QSPR model. Often this issue in QSAR/QSPR studies is addressed in a very heuristic way by experience or by taking an educated guess. However, in general one cannot assume that a descriptor, which is good for one QSAR/QSPR problem will be automatically good for another QSAR/QSPR problem, because similarity of molecules is strongly context dependent (Bender and Glen, 2004). While two molecules can be viewed as highly similar e.g. with respect to their solubility in water, they can be very dissimilar with regard to their oral bioavailability. Hence, it is important to find a description, which captures the features of the molecules in a dataset that are relevant for the QSAR/QSPR problem at hand and this way give rise to a similarity measure that is well suited for the problem.

Another difficulty is that even, if we have a candidate set of promising descriptors, some descriptors can contribute more to the model than others, and it is possible that irrelevant descriptors degrade the performance of our model. Hence, there is the question how we can systematically select a subset of our original descriptors, which is well suited for the model we want to infer.

In general this leads to an instance of the classical feature subset selection problem in machine learning (John et al., 1994; Blum and Langley, 1997; Kohavi and John, 1997; Guyon and Elisseeff, 2003), which in general is known to be NP-complete (Davies and

Russel, 1994). Therefore heuristics have been developed to deal with this problem: In the *filter* approach descriptor selection is performed as a preprocessing step to the actual learning algorithm by means of some predefined relevance measure, which is independent of the actual generalization performance of the learning machine and can thus potentially mislead the selection algorithm (Kohavi and John, 1997). In the *wrapper* approach, on the other hand, the generalization performance of the learning algorithm trained with a given descriptor subset as an input is estimated. The goal is then to look for that descriptor subset that returns the minimal estimated generalization error. Thereby some search heuristic, such as *forward selection* or *backward elimination*, or a stochastic search method, like Genetic Algorithms, is used (Salcedo-Sanz et al., 2002; Vafaie and Jong, 1998; Fröhlich et al., 2004a). Obviously, the wrapper approach requires repeated re-trainings of the underlying learning algorithm and thus is often criticized as being slow. On the other hand it can often give superior results to filters (Guyon et al., 2002; Weston et al., 2001). Besides these two main approaches some learning algorithms have a built-in mechanism of selecting relevant descriptors, like CART decision trees (Breiman et al., 1984), so-called *embedded* methods. Note that the problem of descriptor *selection* should not be confused with the problem of descriptor *construction* as it is e.g. done by Principal Component Analysis (PCA — Hotelling, 1933), where the principal components are linear combinations of the original descriptors.

The descriptor selection problem has also been recognized within the chemoinformatics community for some time (Xue et al., 2004; Byvatov and Schneider, 2004; Agrafiotis and Xu, 2003; Godden et al., 2003; Wegner et al., 2003b). In this Chapter we concentrate on SVM-based descriptor selection techniques only. To our best knowledge SVM-based descriptor selection methods have been rarely known within the context of QSAR/QSPR studies prior to our work (Fröhlich et al., 2004b) and no attempts have been made to investigate the incorporation of expert knowledge into the descriptor selection process. However, in many QSAR/QSPR problems it is known from expert or domain knowledge that certain descriptors must play a role. For instance, it is known that the polar surface area of a molecule is an important factor for the passive transport of a compound through a membrane in the body (Artursson and Bergström, 2003). Thus, a natural question is, whether there is a positive effect from an incorporation of this prior knowledge into the descriptor selection process.

Probably one the earliest publications appearing about the same time as our paper, is that of Xue et al. (2004), who used the Recursive Feature Elimination (RFE) algorithm (Guyon et al., 2002) to select descriptors for the prediction of pharmacokinetic and toxicological properties of chemical compounds. Also about the same time is the paper of Byvatov and Schneider (2004): Utilizing the gradient of the SVM decision function with respect to scaling factors θ of the descriptors, they prune molecular features, which are of low relevance to characterize focused libraries of enzyme inhibitors.

The rest of this Chapter is organized as follows: In Section 5.2 we review some classical feature selection methods, especially for SVMs. After that, in Section 5.3, we explain, how expert domain knowledge can be incorporated into the descriptor selection process. We also propose an algorithm, which based on a prior computed ranking of the descriptors, can lead to a further improvement by means of successive exchange operations between the set of actually selected and not selected descriptors. In Section 5.4 we first perform systematic and extensive evaluations on artificial toy data in order to get a better understanding of both, the effect of bringing prior knowledge into the descriptor selection process as well as our own Rank based Exchange Algorithm. After that, we conduct experiments on real life QSAR/QSPR datasets. We compare the prediction performances achieved by the different algorithms and also investigate, which descriptors are selected by our method. We discuss our results in Section 5.5, before we conclude in Section 5.6.

5.2 Related Methods

5.2.1 Information Theoretic Filter Algorithms

A classical means of filtering descriptors is based on the *mutual information* between a variable X and the target Y (cf. Duda et al., 2001). Let us, by a slight abuse of notation, denote the domain of a variable with the same symbol as its name. Then the mutual information between X and Y is defined as:

$$I(X,Y) = \int_{X \times Y} p(x,y) \log \frac{p(x,y)}{p(x)p(y)} dxdy \tag{5.1}$$

where $p(x)$ and $p(y)$ are the probability densities of x and y, respectively, and $p(x,y)$ is the joint density. The mutual information can be viewed as the Kullback-Leibler divergence between $p(x,y)$ and $p(x)p(y)$, i.e. it measures how close the joint density is related to the product of the individual densities. Note that for $p(x,y) = p(x)p(y)$ the mutual information becomes zero, because then X and Y are statistically independent. Furthermore, the mutual information equals the difference between the *Shannon entropy* $H(X) = \mathbf{E}[-\log p(x)]$ and the *conditional Shannon entropy* $H(X|Y) = \mathbf{E}[-\log p(x|y)]$ (Duda et al., 2001).

An example of an information theoretic filter is the Battiti algorithm (Battiti, 1994). The idea is here to select those descriptors, which have on one hand a high mutual information with regard to the target and on the other hand a low mutual information with regard to all other variables. Thereby the trade-off between both goals can be controlled by a parameter. Closely related to this idea is also the recently published *Fast Correlation Based Filter* (FCBF) algorithm (Yu and Liu, 2004). Here descriptors are selected that have a higher symmetrical uncertainty

$$SU(X,Y) = \frac{2I(X,Y)}{H(X)+H(Y)} \tag{5.2}$$

with regard to the target Y than to any other variable X'. This way not only irrelevant variables are removed, but also redundancy within the selected descriptor subset is minimized.

A problem with information theoretic methods is the estimation of the densities $p(x)$, $p(x, y)$. The simplest and computationally most effective way is the computation of a histogram binning. A more accurate but also much more time consuming solution is the usage of kernel density estimation (e.g. Duda et al., 2001).

5.2.2 SVM-based Wrapper Algorithms

Given the SVM decision hyperplane $\langle \mathbf{w}, \phi(x) \rangle + b$ in feature space, one way of performing descriptor selection is to successively eliminate that descriptor t, for which the absolute value $|w_t|$ of component t of the hyperplane normal vector $\mathbf{w}$ is smallest and which thus has the least influence on the decision hyperplane. This is the basic idea of the Recursive Feature Elimination (RFE) algorithm (Guyon et al., 2002). Unfortunately, the calculation of $|w_t|$ is only possible in the linear case. However, under the assumption that the set of support vectors does not change significantly when eliminating just one descriptor, in the nonlinear case one can look for that descriptor t for which the change of $\|\mathbf{w}\|^2$

$$\Delta\|\mathbf{w}\|^2(t) = \left| \sum_{i,j} \alpha_i^* \alpha_j^* \left(k(\mathbf{x}_i, \mathbf{x}_j) - k(\mathbf{x}_i[-t], \mathbf{x}_j[-t]) \right) \right| \tag{5.3}$$

is smallest. Here $\mathbf{x}[-t]$ denotes that descriptor t has been removed from molecule descriptor vector $\mathbf{x}$ and α^* are the Lagrangian multipliers obtained from the SVM training (c.f. Chapters 3.2, 3.4). Note that a small change of $\|\mathbf{w}\|^2$ also corresponds to a small change of the margin size $1/\|\mathbf{w}\|$. Removing the descriptor that effects $\|\mathbf{w}\|^2$ least therefore is equivalent to removing the descriptor that effects the margin least. In the linear case, this criterion is the same as eliminating the descriptor with smallest $|w_t|$.

It is worth mentioning that the RFE algorithm can also be used in case of a Support Vector Regression. Thereby a small change in (5.3) corresponds to a small change of the ϵ-margin and hence of the smoothness of the function (c.f. Chapter 3.2, 3.4).

As a result of the RFE algorithm one receives a ranking of all descriptors according to their order of elimination. Thereby, for computational reasons, usually half of the existing features is removed at once with respect to (5.3), which leads to $O(\log_2(D))$ SVM retrainings and $O(D \cdot \log_2(D))$ kernel matrix calculations (D is the original number of descriptors).

Instead of computing the absolute difference in the (ϵ-)margin one can also consider the absolute derivative of $\|\mathbf{w}\|^2$ with respect to scaling factors $\theta = (\theta_1, ..., \theta_D)^T \in \mathbb{R}^D$ of the descriptors instead:

$$\left| \frac{\partial \|\mathbf{w}\|^2(\theta)}{\partial \theta_t} \right| = \left| \sum_{i,j} \alpha_i^* \alpha_j^* \frac{\partial k(\mathbf{x}_i * \theta_i, \mathbf{x}_j * \theta_j)}{\partial \theta_t} \right| \tag{5.4}$$

Here "*" denotes a component-wise multiplication of two vectors. This was proposed by Rakotomamonjy (2003) and leads to similar results as the original RFE algorithm. In case of a linear kernel, removing the descriptor t with minimal $\left|\frac{\partial\|\mathbf{w}\|^2(\theta)}{\partial\theta_t}\right|$ is the same as eliminating the one with the smallest $|w_t|$.

It was also suggested to perform gradient descent of either the span bound or the radius-margin error bound (c.f. Chapter 3.2) with respect to scaling factors θ of the descriptors and then to remove those which lead to the smallest scaling factors (Chapelle et al., 2002; Weston et al., 2001). Again, this approach leads to results similar to those of the RFE algorithm, while in general being computationally more demanding (Rakotomamonjy, 2003).

For linear SVMs Weston et al. (2002) proposed an approximate optimization scheme of the zero-norm (i.e. of the number of nonzero components) of the hyperplane normal vector $\mathbf{w}$. Their algorithm effectively works by iteratively training a regular SVM and rescaling each input variable by the corresponding absolute component of $\mathbf{w}$.

The approximate minimization of the zero-norm is also the basic of the algorithm by Bradley and Mangasarian (1998), which requires repeatedly solving a linear program formulation of the SVM and thus makes it rather impracticable for very high dimensional data (such as QSAR/QSPR problems).

Finally, in Krishnapuram et al. (2003) a kernel-based joint classifier and feature selection algorithm is presented that finds the dual coefficients α of a separating hyperplane in feature space (in this case this is not necessarily the maximum margin hyperplane) together with the scaling factors θ of the descriptors. This is achieved in a Bayesian manner by defining Laplacian priors with corresponding hyperpriors for α and θ. Learning is then performed by an EM algorithm (Dempster et al., 1977).

5.3 Descriptor Selection for SVMs using Expert Knowledge and the Rank Based Exchange Algorithm

A straight forward and yet effective method to incorporate expert knowledge into SVM-based descriptor selection is to take the RFE algorithm and explicitly set the measure $\Delta\|\mathbf{w}\|^2(t)$ to ∞, if the corresponding descriptor t has to be in the solution. This leads to an order of the descriptors, in which the known relevant descriptors are automatically ranked highest.

A complementary idea to improve the original RFE solution was presented in Fröhlich and Zell (2004): As the highest ranked descriptors together are not necessarily the optimal ones[1], the idea here is to successively exchange descriptors from the set S of the m

[1]Note that the descriptor selection problem is really a *subset* selection problem. This can lead to the paradoxical situation that is described in Guyon and Elisseeff (2003): A descriptor can be completely useless by itself, but in combination with some others it can *increase* the generalization performance.

highest ranked descriptors with those of the currently removed ones, which form a set R. Both sets can be viewed as a queue. Now we take the first η descriptors from set R and put them in set S. In return we remove those η descriptors from S for which criterion (5.4) is smallest. Thereby, descriptors that are known to be relevant a-priori are not considered. The removed descriptors are then reinserted into R according to the number of times they have been thrown out so far. The whole procedure is repeated, until some termination criterion is met. At each round an estimator of the generalization performance is computed. At the end the solution S^* having the lowest risk bound is returned.

As a risk bound we use the span rule (theorem 3.2.4) for classification and the bound stated in theorem 3.4.1 for regression problems. In the multi-class case we utilize a one-against-rest strategy to compute bounds for each single classifier and then consider their sum as the final score.

The number η of exchanged descriptors can be gradually increased with the number of times the current solution could not improve over the best one in order to eventually capture descriptors, which are lower ranked in R, but are yet useful together with some of those already in S. In pseudo-code 2 we present a detailed description of the whole algorithm, which we call **R**ank Based **E**xchange **A**lgorithm (REA).

In principle it is possible to combine the REA algorithm with arbitrary descriptor rankings. Nonetheless, the combination with RFE seems most natural as the elimination of descriptors from the set S is also guided by a SVM-based criterion. Additionally one should note that REA is a greedy method, which is clearly sensitive to its initialization. Therefore, it should also make sense to combine REA with a ranking provided by RFE with expert knowledge incorporated.

5.4 Experiments

5.4.1 Datasets

Toy Data In order to study the performance of the REA algorithm systematically in a controlled way, we first investigated artificial toy data. Thereby we followed Weston et al. (2001) and Rakotomamonjy (2003), who used this data as a benchmark problem for feature selection algorithms: We created 1000 training and 10000 independent test instances with 202 descriptors. The first six of the 202 features are relevant, but still have inner redundancy. The probabilities of classes $y = +1$ and $y = -1$ are equal. The first three features X_1, X_2, X_3 are drawn as $X_i = y\mathcal{N}(i, 1)$, and the second three features X_4, X_5, X_6 are drawn as $X_i = \mathcal{N}(0, 1)$ with a probability of 0.7. Otherwise the first three features are drawn as $X_i = \mathcal{N}(0, 1)$ and the second three as $X_i = y\mathcal{N}(i - 3, 1)$. The remaining features were drawn randomly from $\mathcal{N}(0, 20)$. The pairwise class separation induced by the first six features is depicted in Fig. 5.1. As one can see the class separation

Algorithm 2 Rank Based Exchange Algorithm

```
Input: ranking queue Q, m
Output: set S* of selected descriptors according to
lowest risk bound
S = first m descriptors from Q
R = removed descriptors
rejected(i) = 0, i = 1, ..., D
bound = ∞
loops = 0
REPAT
   oldbound = bound
   bound = calculated risk bound
   if bound < oldbound
      if improvement > 5%
         notImproved ← 0;
   else
      notImproved ← notImproved + 1
   end
   η = min(2^notImproved, m)
   add first η descriptors from R to S
   take = |S| − m descriptors from S with for which
   criterion (5.4) is smallest
   rejected(take(i)) ← rejected(take(i)) + 1, i = 1, ..., |take|
   set new position of take(i) in R as 2^rejected(take(i))
   R ← R − take
   loops ← loops + 1
UNTIL termination (notImproved == 10)
```

by different pairs of features is quite different. For instance, features 3 and 6 provide a good class separation, whereas that by features 1 and 4 is worse. Generally, a good pairwise classification has to combine one feature from the first and one from the second group of the six relevant ones.

QSPR Datasets Our other datasets are real life QSPR problems: The HIA (Human Intestinal Absorption) dataset consists of 164 structures from different sources. The dataset is a collection of Wessel et al. (1998) (82 structures), Gohlke et al. (2001) (49 structures), Palm et al. (1997) (8 structures), Balon et al. (1999) (11 structures), Kansy et al. (1998) (6 structures), Yazdanian et al. (1998) (6 structures) and Yee (1997) (2 structures). The molecules are divided into two classes "high oral bioavailability" (106 structures) and "low oral bioavailability" (58 structures) based on a histogram binning (Wegner et al., 2003a). Multiple occurrences of one molecule were removed. After the usual elimination of the hydrogen atoms, the maximal molecule size was 57 and the average size 25 atoms. From literature it is known that the polar surface area is a relevant descriptor for this problem (van de Waterbeemd and Gifford, 2003).

The Yoshida dataset (Yoshida and Topliss, 2000) has 265 molecules that we divided into two classes "high bioavailability" (bioavailability $>=$ 50%, 159 structures) and "low bioavailability" (bioavailability $<$ 50%, 106 structures). The maximal molecule size was 36 and the average size 20 atoms, after removing hydrogen. Mandagere and Jones (2003) report the octanol/water partition coefficient logP to be a relevant descriptor.

The BBB (Blood Brain Barrier) dataset (Hou and Xu, 2003) consists of 109 structures having a maximal molecule size of 33 and an average size of 16 atoms after removing hydrogen. The target is to predict the logBB value, which describes up to which degree a drug can cross the blood-brain-barrier. The polar surface area and the octanol/water partition coefficient logP are known to be important for this problem (van de Waterbeemd and Gifford, 2003).

The SOL dataset with 296 molecules was published in Abolmaali et al. (2003) as a test dataset for the SOL project[2]. The dataset consists of four different classes of inhibitors: thrombin inhibitors (75 molecules), serotonin inhibitors of the 5HT2 class (75 molecules), monoamine oxidase inhibitors (71 molecules) and 5-hydroxytryptamine oxidase (75 molecules). The goal is to learn the classification of the structures into these four categories. After removing hydrogen atoms, the maximal molecules size was 48 and the average 28 atoms. There are no descriptors, which are a-priori known to be relevant for this problem.

Finally, we investigated the Huuskonen dataset (Huuskonen, 2000), which has 1264 molecules with a maximal size of 47 and an average size of 13 atoms after removing hydrogen. The goal is to predict the aqueous solubility of a structure measured by the

[2]Search and Optimization of Lead Structures (SOL), German Federal Ministry of Education and Research (bmb+f), contract no. 311681

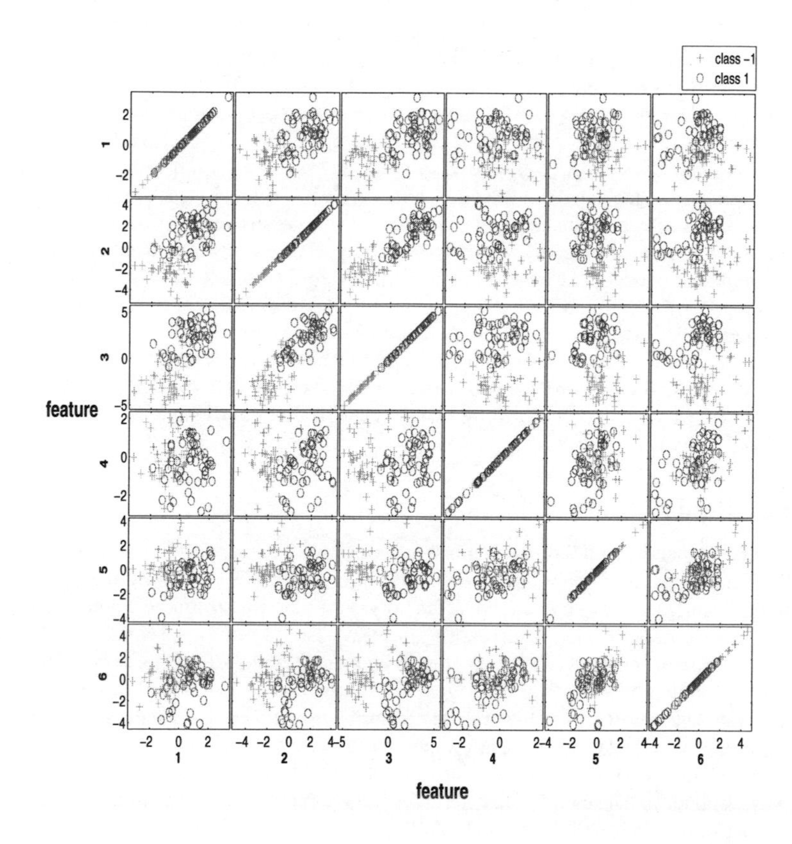

Figure 5.1: Toy data: Separation induced by pairwise combinations of features 1,...,6. Blue crosses indicate points belonging to class -1 and red circles those belonging to class +1.

logS value. An important known factor for this is the octanol/water partition coefficient logP (Norinder and Haeberlein, 2003).

For each dataset we calculated all descriptors available in the open source software JOELib[3] developed in our group (Wegner, 2006). This way for the HIA dataset we obtained 6603, for the Yoshida dataset 5857, for the BBB dataset 5456, for the SOL dataset 5774, and for the Huuskonen dataset 6172 descriptors. The difference is due to the removal of constant descriptors. JOELib contains various 0D - 3D descriptors[4]. Besides others, these include the Radial Distribution Function descriptor, the Moreau-Broto autocorrelation, the Global Topological Charge Index and Burden's Modified Eigenvalues (Todeschini and Consonni, 2000). The descriptors are based on the following atom properties: atom mass (tabulated), valence (calculated, based on graph connectivity), conjugated environment (calculated, SMARTS based), van-der-Waals volume (tabulated), electron affinity (tabulated), electro-negativity (tabulated, Pauling), graph potentials (calculated, graph theoretical), Gasteiger-Marsili partial charges (calculated, iterative — Gasteiger and Marsili, 1978), intrinsic state (calculated), electrotopological state (calculated), electrogeometrical state (calculated — Todeschini and Consonni, 2000).

Each dataset consists of energy-minimized structures using the MOE all-atom-pair force field method (Halgren, 1998), and was tested for duplicate molecules. Missing values in descriptors were replaced by mean values, which corresponds to a maximum likelihood estimate.

5.4.2 Results

Toy Data: Experimental Setup In a first experiment we compared a linear hard margin SVM trained with all features, the standard RFE algorithm, the RFE algorithm with some expert information incorporated (RFE+EXP), the Fast Correlation Based Filter (FCBF) algorithm and the REA method on the toy problem described above. Thereby the REA method relied either on the ranking previously computed by the RFE algorithm or on the ranking computed by the RFE+EXP algorithm. In the latter case we call this approach REA+EXP. The expert knowledge was that feature 6 is relevant. For the density estimations necessary in the FCBF algorithm we used a simple histogram binning with $n/\log n$ bins (n = number of training patterns), because of the high computational effort of kernel density estimation. The cutoff parameter δ for the symmetrical uncertainty below which features are not considered was set to 0.

We sampled uniform randomly 10, 20, 30, 40, 50, 70 and 100 training points out of the set of 1000 training points without replacement and measured the test error on the 10,000 independent test points. In the training set all features were normalized to mean

[3] http://sourceforge.net/projects/joelib

[4] A complete list of all actually implemented descriptors can be found under http://wiki.cubic.uni-koeln.de/bowiki/index.php/JOELib/AlgorithmDictionary.

0 and standard deviation 1 and the resulting scaling factors then applied to normalize the test data. The whole procedure was repeated 10 times for each number of training points.

Toy Data: Evaluation The results on the toy data (Fig. 5.2 and 5.3) show that the REA+EXP algorithm constantly leads to the lowest error rates, closely followed by the RFE+EXP model. Both error curves are significantly different at the 10% level from those of the original REA and RFE methods according to the Wilcoxon signed rank test ($p = 4.69\%$, 1.56%). This reveals the usefulness of incorporating the expert knowledge. Moreover, the REA method significantly outperforms RFE ($p = 6.25\%$) without the usage of expert knowledge. Additionally, REA was **not** significantly worse than the RFE+EXP model. Both results indicate a positive effect of the REA algorithm. The FCBF method was outperformed significantly by all wrappers ($p < 4\%$), which highlights the weakness of a filter approach. Thereby the FCBF algorithm on average selected 3 ± 1 features for 10 data points and 1 ± 3 features in all other cases. All feature selection algorithms achieved a significant improvement compared to the SVM model using all features ($p < 5\%$).

We further analyzed the frequency, with which the individual features on average were selected by the different algorithms, depending on the number of training instances (Fig. 5.4 - 5.7): While the standard RFE algorithm in all cases failed to select one feature from the first group and one from the second group of the six relevant features, the REA method showed a much better behavior. Except for 20 and 100 training points in all cases the feature combinations 3 and 6 or 3 and 5 were selected most. The incorporation of expert knowledge for both, RFE and REA, had a positive effect. Feature 3 this way was selected very frequently from the beginning on. Thereby, there was almost no difference in the behavior of RFE+EXP compared to REA+EXP.

In Fig. 5.8 we depict the number of loops needed until termination for REA and REA+EXP. The graphs in all cases reveal a moderate computational demand. Furthermore, there is a slight increase of the running time in dependency of the number of training points. This results from the fact that with a very small number of training points no accurate estimation of the generalization performance of the SVM via the span rule (theorem 3.2.4) is possible, and hence the bound often equals zero. If the bound is zero, then obviously no improvement can be achieved, and therefore the algorithm terminates. With a higher number of training instances, this scenario occurs less frequently. From Fig. 5.8 we can also see that incorporating expert knowledge slightly reduces the running time, because a local minimum of the error bound can be found earlier.

Real Life Data: Experimental Setup Let us now come to our experiments performed on our QSPR real-life datasets. We evaluated the same descriptor selection algorithms as before, but this time also included a model purely based on the expert provided descriptors (EXP) . We used 10-fold cross-validation to assess the prediction quality. We employed a SVM with RBF kernel on the classification datasets (HIA , Yoshida and SOL)

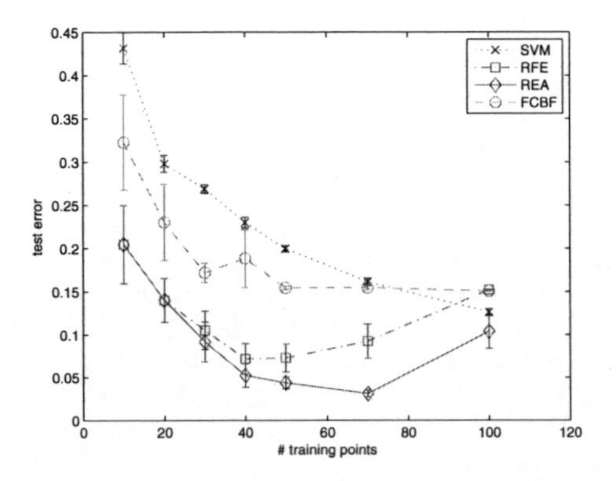

Figure 5.2: Toy data. Average test error (± standard error) over 10 trials in dependency of the number of training points: REA vs RFE and FCBF.

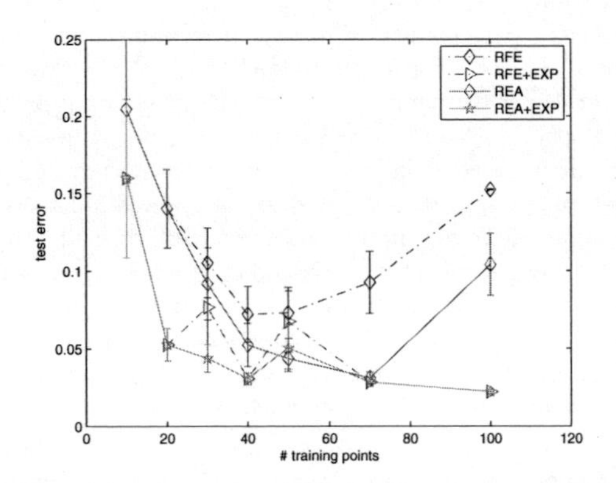

Figure 5.3: Toy data. Average test error (± standard error) over 10 trials in dependency of the number of training points: REA vs RFE: Effect of incorporating expert knowledge.

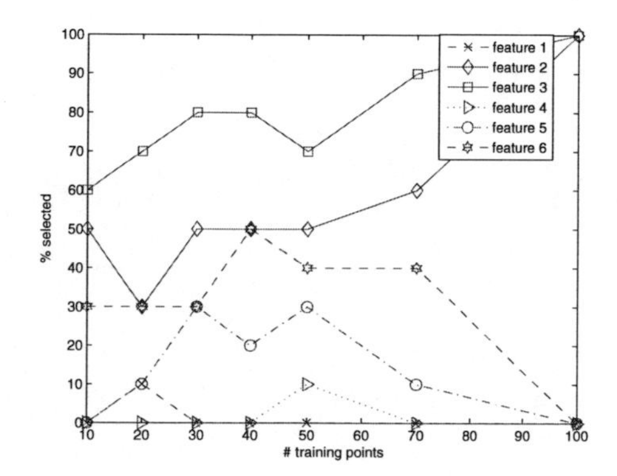

Figure 5.4: Toy data. Average selection of the first six features depending on the number of training points: RFE algorithm.

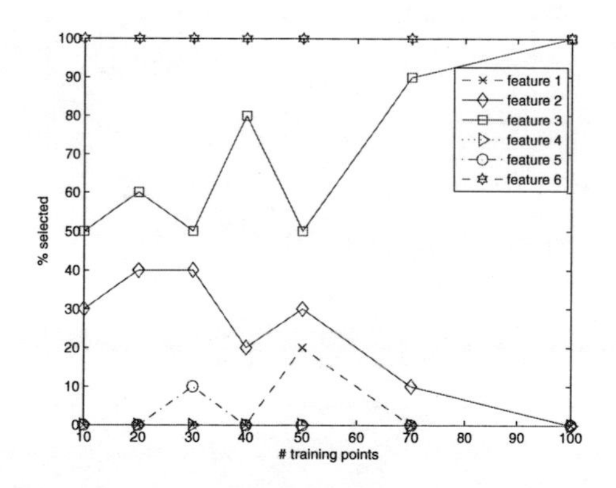

Figure 5.5: Toy data. Average selection of the first six features depending on the number of training points: RFE+EXP algorithm.

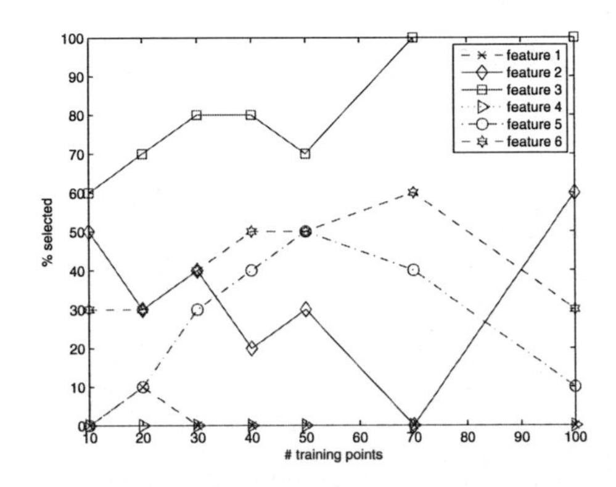

Figure 5.6: Toy data. Average selection of the first six features depending on the number of training points: REA algorithm.

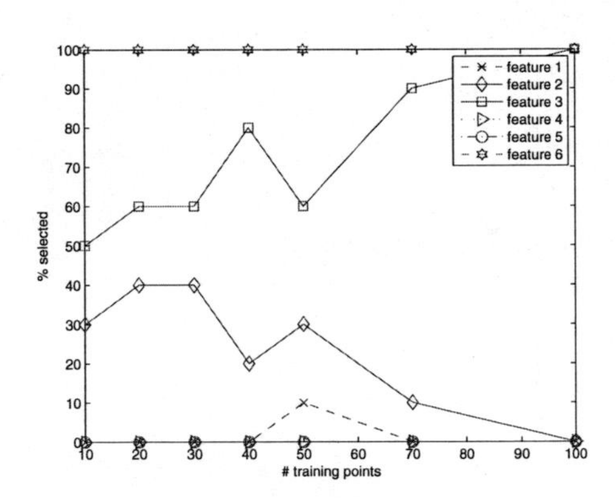

Figure 5.7: Toy data. Average selection of the first six features depending on the number of training points: REA+EXP algorithm.

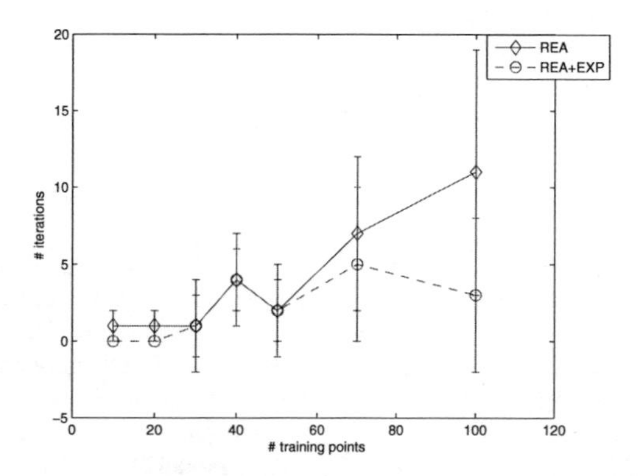

Figure 5.8: Toy data. Average number of iterations ($\pm$ standard deviation) needed until termination by REA and REA+EXP.

and a ϵ-Support Vector Regression (SVR) on the regression datasets (BBB and Huuskonen). On the classification problems we ensured that the ratio of examples from both classes in the actual training set was always the same (stratified cross-validation). On each actual training fold a model selection for the necessary parameters was performed by evaluating each candidate parameter set by an extra level of 10-fold cross-validation. This included the choice of the soft-margin parameter C from the interval $[2^{-2}, 2^{14}]$, and on the regression datasets (BBB, Huuskonen) additionally the width of the ϵ-tube from the interval $[2^{-8}, 2^{-1}]$. The width of the RBF kernel was tuned in the range $\sigma/4, ..., 4\sigma$, where σ was set such that $\exp(-D/(2\sigma^2)) = 0.1$ (D = number of descriptors). Concerning the RFE algorithm, after each removal of half of the existing descriptors the width σ was adapted by multiplying it with $\frac{1}{\sqrt{2}}$[5]. The number of descriptors for the RFE algorithm was determined by evaluating the 5-fold cross-validation error after each removal step. To be better comparable, for the REA methods the same number of descriptors as for the RFE algorithms was used. All descriptor values (also the target values logBB and logS in case of the regression datasets) were normalized to mean 0 and standard deviation 1 on each training fold, and the calculated scaling factors were then applied to normalize the descriptor values in the actual test set.

[5]This is because we have $\sigma = \sqrt{-D/(2 * \log(0.1))}$ and hence for $D' = D/2$ $\frac{\sigma'}{\sigma} = \frac{\sqrt{0.5D/(2*\log(0.1))}}{\sqrt{D/(2*\log(0.1))}} = \frac{1}{\sqrt{2}}$.

Real Life Data: Evaluation The results are shown in in Fig. 5.9 - 5.15 and Tables D.1 - D.2 in the Appendix. For the regression problems, besides the mean squared error (MSE) we also computed the squared Pearson correlation (r^2) between the predicted values, $\hat{\mathbf{y}}$, and the true values, $\mathbf{y}$, which is an often used performance measure in the chemoinformatics literature:

$$r^2 = \left(\frac{\sum_{i=1}^{n} (y_i - \bar{y})(\hat{y}_i - \bar{\hat{y}})}{(n-1) \cdot s_{\hat{\mathbf{y}}} \cdot s_{\mathbf{y}}} \right)^2 \tag{5.5}$$

Here $s_{\hat{\mathbf{y}}}$ and $s_{\mathbf{y}}$ denote the standard deviations of $\hat{\mathbf{y}}$ and $\mathbf{y}$, respectively, and $\bar{y}$ and $\bar{\hat{y}}$ their means.

All SVM-based descriptor selection methods could significantly improve over the full descriptor model (SVM/SVR) on the SOL, the BBB and the Huuskonen dataset. Thereby statistical significance was tested by a two-tailed paired t-test at the 10% significance level. The difference of RFE+EXP to the DESC model on the BBB dataset is regarded as significant according to the squared correlation measure ($p = 8.97\%$) only, whereas the other differences are significant to both, MSE and r^2. The FCBF algorithm was significantly **worse** than the full descriptor model on the SOL dataset ($p = 0.16\%$) and the Huuskonen dataset ($p < 0.1\%$ for MSE/r^2). Furthermore, FCBF was significantly outperformed by the other descriptor selection algorithms on the SOL, the Huuskonen dataset ($p < 0.1\%$ for MSE/r^2) and on the BBB dataset. Incorporating expert knowledge into the RFE algorithm lead to no significant improvements of the error rate, but on the HIA and the BBB dataset fewer features were selected. The REA method yielded comparable results to RFE. However, REA+EXP significantly improved over RFE on the BBB dataset ($p = 9.03\%$ for r^2), whereas RFE+EXP did not. Moreover, on the BBB dataset REA+EXP was significantly better than the full descriptor according to MSE and squared correlation measure ($p = 7.2\%$ for MSE, $p = 5.3\%$ for r^2), whereas for RFE+EXP the difference to DESC was found significant only for the squared correlation ($p = 8.97\%$).

Applying the REA algorithm in all cases lead to a high improvement of the generalization error bound compared to the original ranking, which suggests a higher statistical stability (Fig. 5.16, Table D.3 in the Appendix). Both results indicate a positive effect of the REA algorithm. At the same time only a few loops were needed until the termination criterion was met (Fig. 5.17). Interestingly enough, on the HIA and the BBB dataset the EXP models obtained the lowest error rates. On the BBB dataset the differences to RFE and REA are significant ($p = 5.13\%, p = 4.25\%$ for r^2).

We next investigated, which descriptors were selected by the REA method on **all** training runs during the cross-validation procedure and thus can be regarded as highly relevant for the chemical problem, because they are detected in a statistical stable way. On the HIA dataset this was the polar surface are (PSA), which is consistent with the expert model. Likewise, on the BBB dataset the PSA and the logP descriptors were selected and on the Yoshida dataset only the logP descriptor. On the SOL dataset Burden's modified

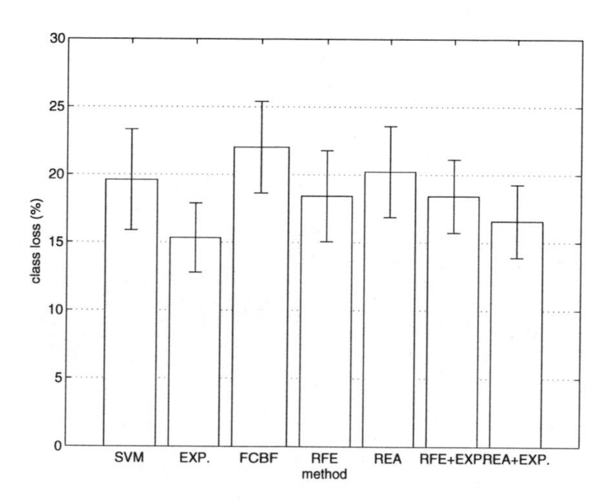

Figure 5.9: 10-fold cross-validation result (class loss in %) for the HIA dataset.

eigenvalues with regard to the electrotopological and the electrogeometrical state were chosen. For the Huuskonen dataset the full list of descriptors is shown in Table D.4 in the Appendix. It includes the logP descriptor, the PSA, the geometrical radius of the molecule, Burden's modified eigenvalues with regard to Pauling's electronegativity and the partial charge, and several descriptors related to the occurrence of H-bond donors and H-bond acceptors. Interestingly, features describing the last two properties can also be found in the model reported by Huuskonen (2000).

We would like to emphasize that the descriptors consistently selected on all training runs are not necessarily the only relevant ones from a chemical point of view. However, giving a comprehensive list is not the primary subject of this study, where we focus on the principle usefulness of algorithms for descriptor selection and want to get insights into their behavior.

Comparison to Non-SVM Method We conducted a comparison of our SVM-based models (SVM without descriptor selection, SVM+FCBF descriptor selection, SVM+RFE, SVM+REA) to a CART tree (Breiman et al., 1984), which is classical and widely used machine learning method for both, classification and regression tasks. CART trees belong to the class of embedded descriptor selection methods, where the choice of the relevant descriptors is part of the learning process itself.

The results of our comparisons are shown in Fig. 5.18 - 5.24 and Tables D.1 - D.2 in the Appendix. The CART tree is significantly outperformed by both, RFE and REA, on

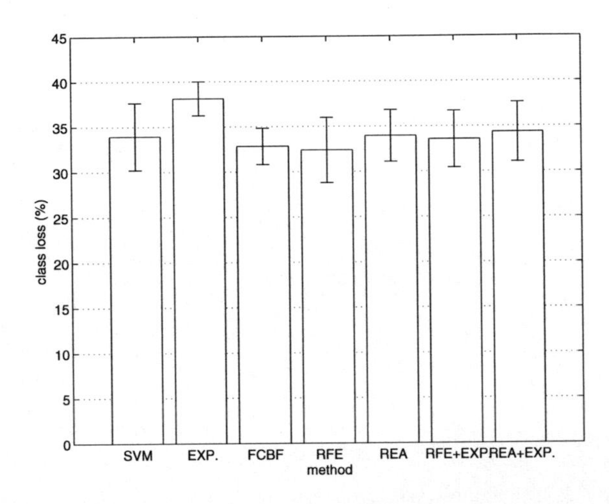

Figure 5.10: 10-fold cross-validation result (class loss in %) for the Yoshida dataset.

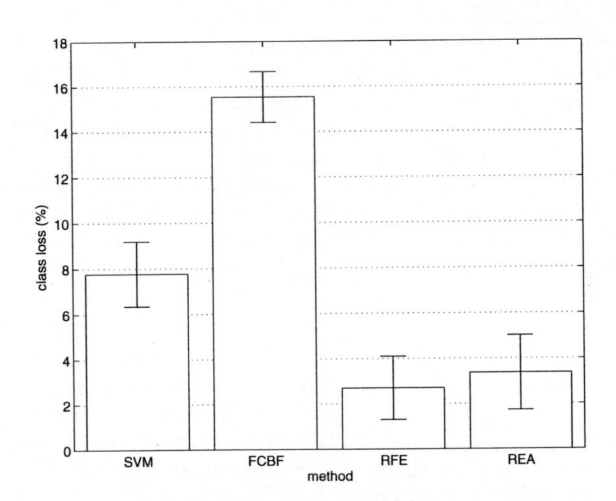

Figure 5.11: 10-fold cross-validation result (class loss in %) for the SOL dataset.

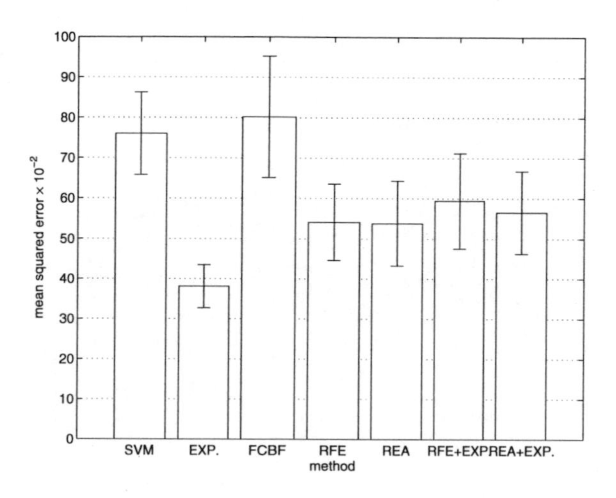

Figure 5.12: 10-fold cross-validation result (mean squared error) for the BBB dataset.

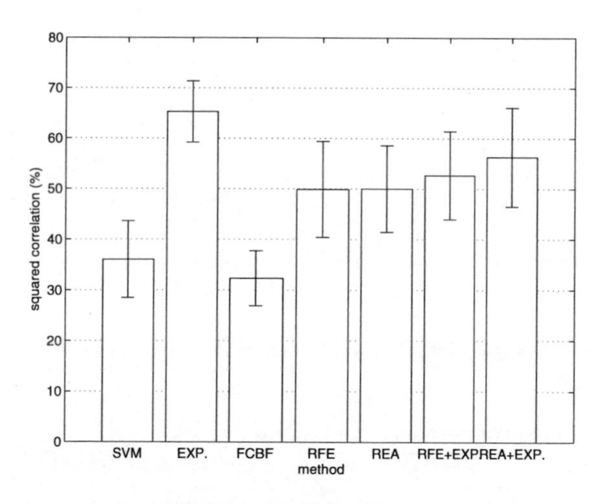

Figure 5.13: 10-fold cross-validation result (squared correlation in %) for the BBB dataset.

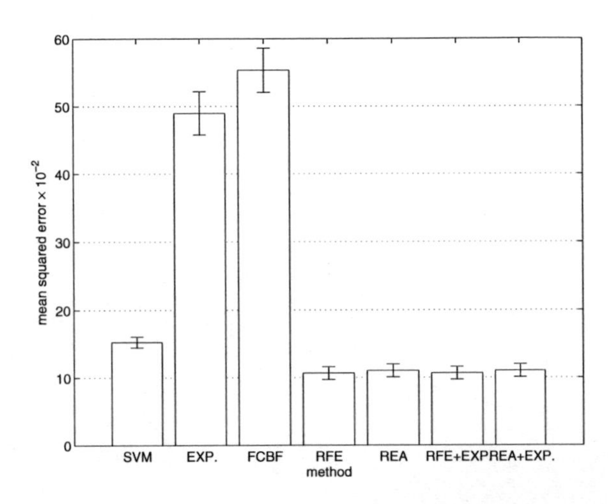

Figure 5.14: 10-fold cross-validation result (mean squared error) for the Huuskonen dataset.

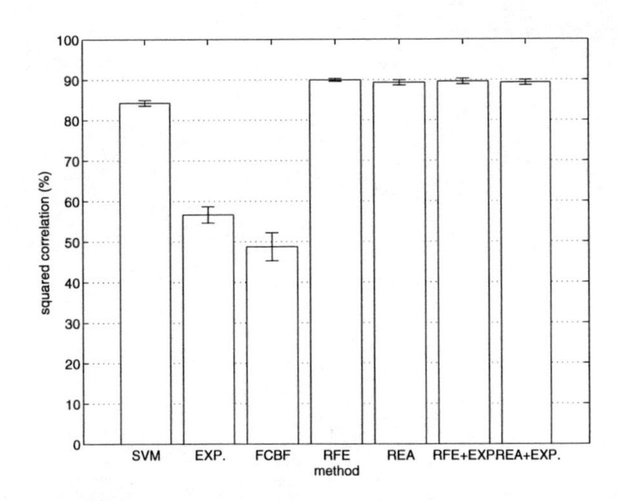

Figure 5.15: 10-fold cross-validation result (squared correlation in %) for the Huuskonen dataset.

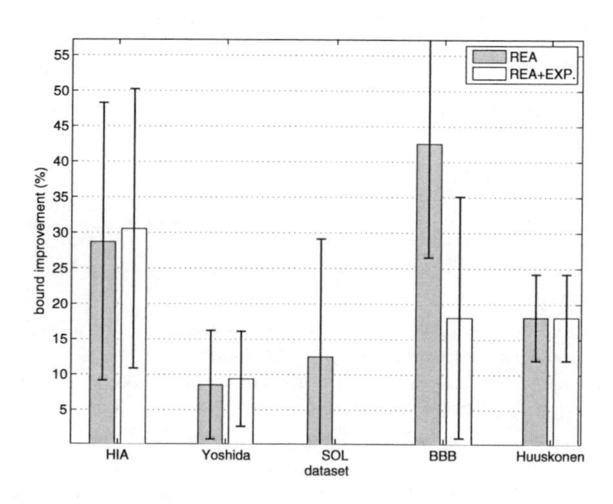

Figure 5.16: REA algorithm: average improvement in % on generalization error bound.

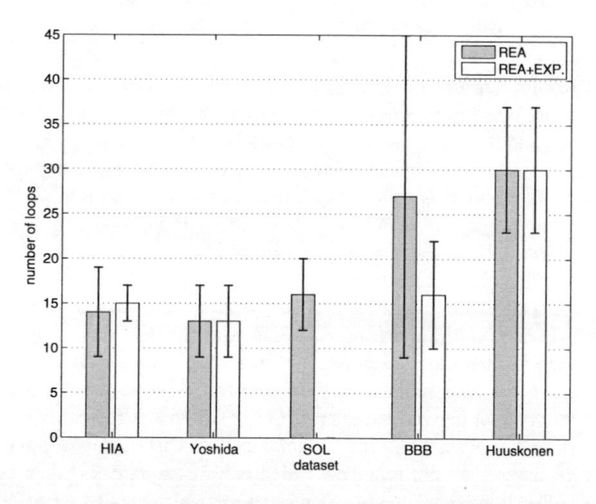

Figure 5.17: REA algorithm: number of loops needed until termination.

Table 5.1: Number of selected features by the different methods ($\pm$std. dev.).

Method	HIA	Yoshida	SOL	BBB	Huuskonen
EXP.	1	1	–	2	1
FCBF	8 ± 1	13 ± 3	7 ± 1	9 ± 2	9 ± 2
RFE	760 ± 2005	977 ± 1931	141 ± 91	183 ± 418	343 ± 201
REA	760 ± 2005	977 ± 1931	141 ± 91	183 ± 418	343 ± 201
RFE+EXP.	104 ± 112	1195 ± 1871	–	28 ± 33	343 ± 201
REA+EXP.	104 ± 112	1195 ± 1871	–	28 ± 33	343 ± 201
CART	16 ± 1	43 ± 3	24 ± 2	35 ± 6	411 ± 7

the SOL ($p = 0.51\%, p = 1.42\%$) and the Huuskonen dataset ($p < 0.1\%$ for MSE/r^2). It is further significantly worse than RFE on the Yoshida dataset ($p = 5.19\%$). Additionally, on the Huuskonen dataset the difference of the CART tree to the full descriptor model is significant ($p < 1\%$ for MSE/r^2).

Interestingly, the number of descriptors selected by the CART tree on the HIA, Yoshida and SOL dataset on average is much lower than for RFE and REA (Table 5.1). Also the variance of the number of selected descriptors is much below that observed for RFE and REA.

In comparison to the FCBF algorithm the CART tree performs significantly better on the SOL ($p = 1.96\%$), the BBB ($p = 9.10\%$ for r^2) and the Huuskonen dataset ($p < 0.1\%$ for MSE/r^2), which underlines the weakness of the filter approach.

Literature Results A direct comparison of our classification/regression results to others from literature is rather problematic, since first, not the same system to calculate descriptors is used and therefore the set of potentially available descriptors is not identical, second, the model is often evaluated using a single splitting into training and test set only, third, the learning problem is sometimes formulated with a different number of classes or as a regression problem instead of a classification problem, and fourth, not the same learning algorithm to build the model is employed. We thus report these results just for the sake of completeness in Table 5.2.

Wessel et al. (1998) treat the HIA problem as a regression task. A neural network is trained on 67 out of 86 molecules using 127 diverse descriptors that were selected using a correlation based feature selection strategy. Nine molecules are used as a validation set to determine stopping of the neural network training and ten structures form an independent test set, for which a root mean squared error (RMS) of 16% is reported.

In (Yoshida and Topliss, 2000) the Yoshida dataset is handled as a four class problem. Doublets in the dataset are not removed. Adaptive least squares is taken as the learning algorithm, which is trained on a set of 232 molecules represented by 18 descriptors, which were chosen a-priori. These include the log distribution coefficient log D and several

Figure 5.18: Comparison of SVM-based descriptor selection models versus a CART tree: 10-fold cross-validation results for the HIA dataset (class loss in %).

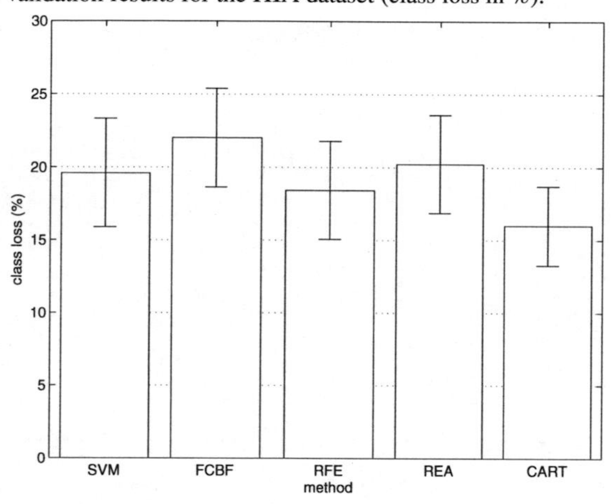

Figure 5.19: Comparison of SVM-based descriptor selection models versus a CART tree: 10-fold cross-validation results for the Yoshida dataset (class loss in %).

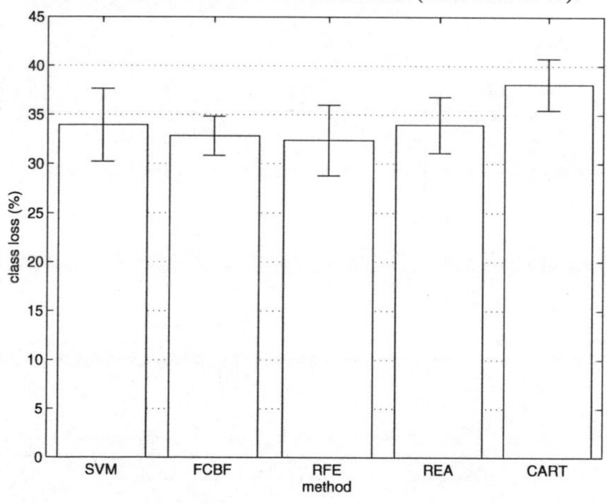

Figure 5.20: Comparison of SVM-based descriptor selection models versus a CART tree: 10-fold cross-validation results for the SOL dataset (class loss in %).

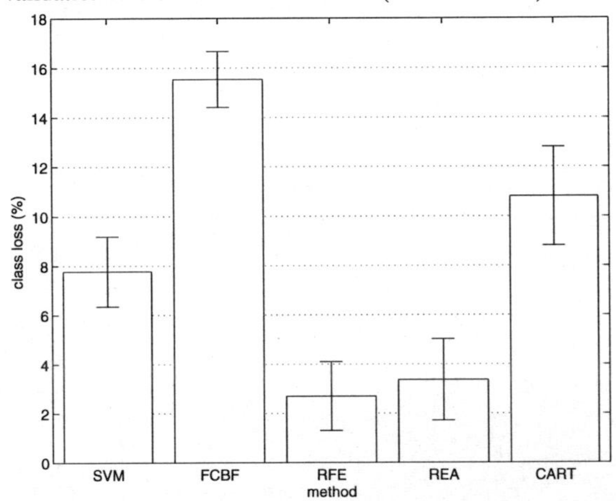

Figure 5.21: Comparison of SVM-based descriptor selection models versus a CART tree: 10-fold cross-validation results for the BBB dataset (mean squared error).

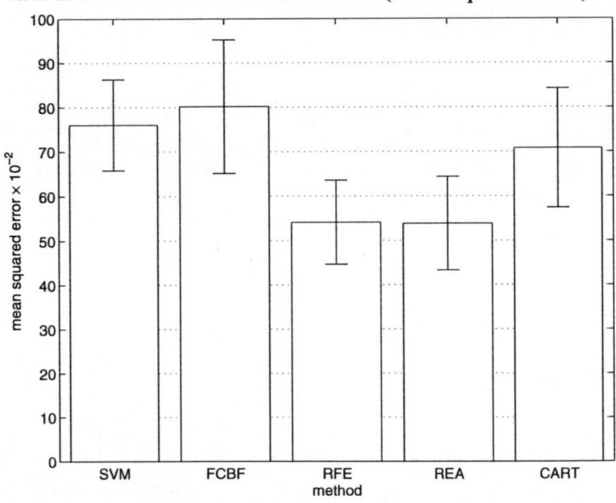

Figure 5.22: Comparison of SVM-based descriptor selection models versus a CART tree: 10-fold cross-validation results for the BBB dataset (squared correlation in %).

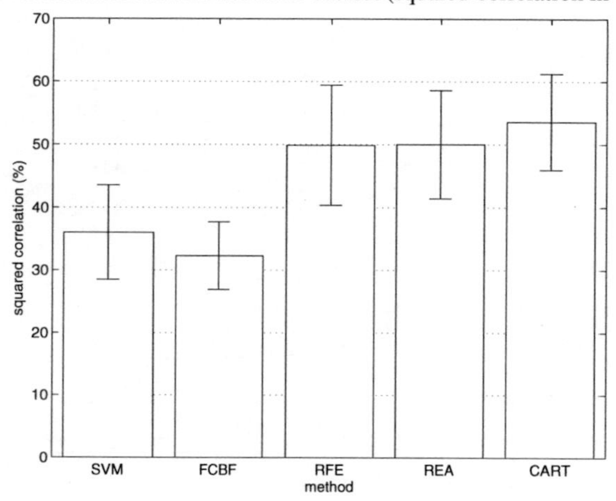

Figure 5.23: Comparison of SVM-based descriptor selection models versus a CART tree: 10-fold cross-validation results for the Huuskonen dataset (mean squared error).

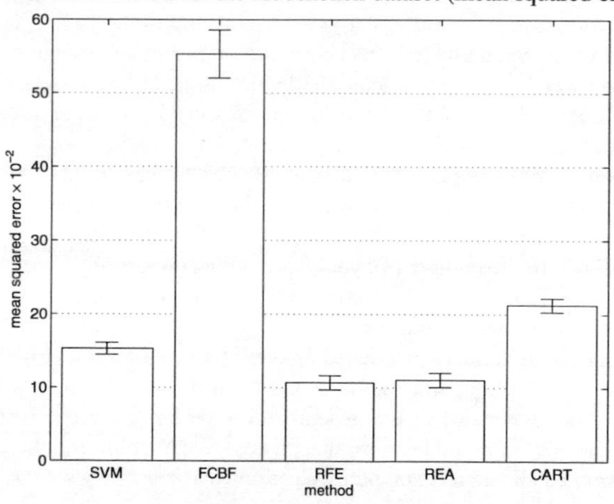

Figure 5.24: Comparison of SVM-based descriptor selection models versus a CART tree: 10-fold cross-validation results for the Huuskonen dataset (squared correlation in %).

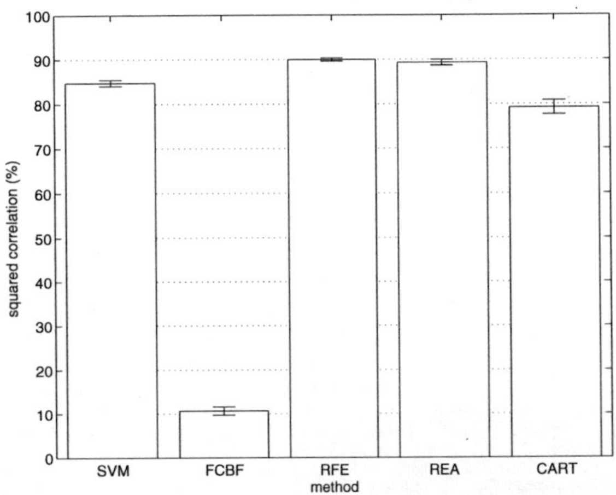

structural descriptors related to metabolism. The evaluation is done on a separate test set of 40 structures and yields a class loss of around 40%.

On the BBB dataset (Hou and Xu, 2003) the authors use a combined multiple linear regression and spline model trained on 78 molecules using modified logP and polar surface area descriptors and the molecular weight. Doublets in the dataset are not removed. The evaluation is done on two test sets consisting of 14 and 23 compounds, respectively, where only the second one was structurally diverse as the test sets used in our evaluation procedure. Hence, we only show the result on the second test set in Table 5.2, which is comparable to that of our EXP and REA+EXP models.

Huuskonen (2000) applies a neural network, which was trained on 884 molecules using 55 a-priori chosen descriptors. These include topological indices, shape indices, valence molecular connectivity indices, atom-type electrotopological state indices, and descriptors related to the number of hydrogen-bonding donors and acceptors. The evaluation is done on a test set of 413 structures yielding a result not very different from our SVM descriptor selection based models.

Apart from the difficulties mentioned above, in all cases a direct comparison to our results is very problematic, since none of the cited methods used cross-validation and hence the reported estimates of the generalization performance are much less reliable than ours. Especially it should be noted that on the HIA, Yoshida and the BBB dataset the test sets are very small and it is not totally clear on what basis they were actually selected.

Table 5.2: Results for our datasets known from literature.

	HIA (Wessel et *al.*)	**Yoshida**	**BBB**	**Huuskonen**
method	neural network	adapt. least squares	mult. lin. regr.	neural network
regr./classif.	regr.	classif. (4 classes)	regr.	regr.
# descriptors	127	18	3	55
size training/test set	67/10	232/40	78/23	884/413
cross-valid. yes/no	no	no	no	no
test error/CV error	16% (RMS)	40% (class loss)	62.41% (r^2)	92.00%(r^2)

5.5 Discussion

In the previous experiments we systematically investigated the following points:

1. The effect of descriptor selection with regard to the generalization performance of a SVM-based QSAR/QSPR model.
2. The performance of SVM-based descriptor selection in comparison to a modern filter approach.
3. A comparison of a SVM-based descriptor selection versus a CART tree model, which has a build-in descriptor selection mechanism.
4. The effect of incorporating expert knowledge into the SVM-based descriptor selection process.
5. The usefulness of the REA algorithm.

Our experiments yielded the following results concerning these questions:

1. A systematic descriptor selection seems to be mandatory in QSAR/QSPR studies, because by means of a SVM-based descriptor selection method in all cases we could achieve significant improvements compared to the usage of all descriptors.
2. Our experiments on toy data as well as on real life data clearly demonstrated the superiority of SVM-based descriptor selection methods compared to the FCBF algorithm. FCBF in several cases was significantly outperformed.
3. The CART tree was significantly outperformed by the SVM models is several cases. In conclusion it appears that SVM-based descriptor selection techniques offer a higher statistical stability and robustness.

4. Although on real life data prior knowledge of relevant descriptors did not help to improve the prediction performance, our experiments on artificial toy data revealed the principle usefulness of such an approach. The reason, why no improvement could be observed on our QSPR datasets, was that the relevant descriptors had been selected by the standard versions of the RFE or REA algorithm anyway.

5. We could show that the REA algorithm on artificial toy data as well as on real life data had a positive effect. While on one hand the cross-validation errors on our QSPR datasets were comparable to those by RFE, on the other hand there was a strong improvement of the theoretical error bounds, which we computed before and after applying our method. Our studies on artificial toy data further underline the principle usefulness of the REA method. Here we could demonstrate a significant improvement compared to RFE and a better selection behavior of the relevant features.

5.6 Conclusion

In this Chapter we dealt with the important problem of descriptor selection in QSAR/-QSPR studies. For this purpose we systematically compared SVM-based QSAR/QSPR with and without descriptor selection and found a systematic descriptor selection to be mandatory. We further showed that SVM-based descriptor selection techniques, like the RFE algorithm, yielded good results on all our problems and are significantly superior to an information based filter approach and a CART tree. We also studied the effect of bringing prior knowledge on relevant descriptors into the descriptor selection process. The usefulness of such a procedure was confirmed by our experiments on artificial toy data. On our real life datasets the relevant descriptors were found, even if no "hint" was given to the algorithm. This does not, however, imply that incorporating expert knowledge in principle is dispensable for QSAR/QSPR problems. Much more, it just tells us that finding the relevant descriptors on our datasets was simple enough without any prior knowledge. Nevertheless, in general it seems a good idea to use as much prior knowledge on a problem as possible in order to improve the prediction performance of the SVM model. To achieve this goal different approaches from the one presented here may be possible as well.

We proposed the Rank based Exchange Algorithm as a greedy method to improve the statistical stability of the solution computed by the RFE algorithm. We studied the principle usefulness of this method on artificial toy data, where we could show that it can improve the feature selection behavior and the prediction performance. Our results on QSAR/QSPR datasets showed a strong enhancement of theoretical error bounds. In conclusion, the REA algorithm seems to have a positive effect. At the same time the extra amount of computation time needed is affordable.

We applied our method to search for descriptors that were consistently selected during the cross-validation procedure on our QSAR/QSPR datasets and found the results to be supported by the literature in those cases, for which the corresponding information was available.

All in all we think that SVM-based techniques provide a reliable and systematic approach to the problem of descriptor selection and can thus help to obtain better QSAR/-QSPR models in the future.

Chapter 6

Kernel Functions for Attributed Molecular Graphs

6.1 Introduction

In the previous Chapter we dealt with a classical descriptor-based representation of molecular structures. The problem we had to face in this context was to find out, which molecular descriptors are actually best suited for the QSAR/QSPR task at hand. This descriptor selection process, which follows the expensive calculation of a huge number of descriptors, is a difficult and computationally demanding process. Additionally, it is not always easy to interpret the selected descriptors in a simple way, like Burden's eigenvalues (Todeschini and Consonni, 2000).

Hence, an appealing idea is to directly work on a representation of chemical compounds as multiple labeled graphs without explicitly calculating any descriptor information. An advantage of this method is that the problem of descriptor selection becomes almost irrelevant, because all computations are carried out directly on the molecular structures represented as labeled graphs. Atoms in a molecule are represented as nodes in the graph and bonds as edges between nodes. Each atom and each bond has certain physico-chemical properties. These properties can be represented as labels of the nodes and edges, respectively. It is also possible to encode structural aspects into the labels, like the membership of an atom to a ring, to a donor, an acceptor, etc. The graph representation can give us a detailed description of the topology of a molecule without making any a-priori assumptions on the relevance of certain chemical descriptors for the whole molecule. It is clear that thereby a crucial point is to capture the characteristics of each single atom and bond by its physico-chemical properties – e.g. electrotopological state (Todeschini and Consonni, 2000), partial charge (Gasteiger and Marsili, 1978) – which are encoded in the labels (see experimental Section for more detail). This representation opens the opportunity to use graph mining methods (Washio and Motoda, 2003) to deal with chemical graph structures. Thereby a principle question is how different graph structures can be compared.

One way of doing so is to define a positive semidefinite and symmetric kernel function (c.f. Chapter 2.5), which measures the degree of similarity between two molecular graphs. In Kashima et al. (2003) the authors propose a kernel function between labeled graphs, which they call *marginalized graph kernel*: Its idea is to compute the expected match of all pairs of random walk label sequences up to infinite length. An efficient computation can be carried out in a time complexity proportional to the product of the size of both graphs by solving a system of linear simultaneous equations. Kashima et al. show that also the geometric and the exponential graph kernel (Gärtner et al., 2003) can be seen as special variants of the marginalized graph kernel (Kashima et al., 2004).

In contrast, the *pattern-discovery* (PD) method by Raedt and Kramer (2001) counts the set of all label sequences, which appear in more than p graphs with p being a so called *minimum support* parameter. Furthermore, it is possible to add extra conditions, for example selecting only paths frequent in one class and scarce in another class. The PD method was especially designed for predicting toxicity of molecules, which from a chemical viewpoint mainly depends on the presence or absence of certain functional groups, and achieves about the same excellent performance there as the marginalized graph kernel (Helma et al., 2001; Kashima et al., 2003, 2004).

Within the chemoinformatics community graph representations, also called *coordinate-free* encodings (Maggiora and Shanmugasundaram, 2004), of molecular structures have been studied as well: Examples thereof are *atom-pair descriptors* (Carhart et al., 1985) and *feature trees* (Rarey and Dixon, 1998). *Maximum common substructure* approaches (Raymond et al., 2002) have been used to calculate similarity measures for chemical graphs. However, all of these approaches differ from the marginalized graph kernel in the way that they do not define a positive semidefinite and symmetric kernel function and hence cannot be used in combination with a kernel-based learning algorithm. Hence, they are not considered further within this work.

A drawback of marginalized graph kernels, when applied to the comparison of molecular graphs, is that they do not have a clear interpretation from a chemists point of view. Thus the goal of our work is to define a valid kernel function for chemical compounds, which, like the marginalized graph kernel, can be used in combination with a SVM or SVR to infer QSAR/QSPR models for various tasks, but better reflects a chemist's intuition on the similarity of molecular graphs. Rather than comparing label sequences, the main idea in our approach (c.f. Fröhlich et al., 2005c,b, 2006b) is that the similarity of two molecular graphs mainly depends on the matching of certain substructures, e.g. aromatic rings, carbonyl groups, etc., and the way they are connected (Fig. 6.1). I.e. two molecules are more similar the better structural elements from both graphs fit together and the more these structural elements are connected in a similar way. Thereby the physico-chemical properties of each single atom and bond in both structures have to be considered.

On an atomic level this leads to the idea of computing a maximum weighted bipartite matching (*optimal assignment*) of atoms in one structure to those in another one, including for each atom information on the neighborhood and on its physico-chemical properties (Fig. 6.2). The maximum weighted bipartite matching is a classical problem from graph theory and can be computed efficiently in cubic time complexity (Mehlhorn and Näher, 1999) and allows an easy interpretation from the chemistry side. To our best knowledge neither similarity measures for molecules based on optimal assignments have been investigated so far, nor attempts to convert it into a proper kernel function, as shown in the following of this Chapter.

A natural extension of our approach is to represent each molecule not on an atomic level, but in form of a *reduced graph*. Thereby certain structural motifs, like rings, donors, acceptors, are collapsed into one node of the molecular graph, whereas remaining atoms are removed. In the literature this procedure is called *pharmacophore mapping* (Martin, 1998). Furthermore, we investigate the effect of combining descriptor information, which is a-priori known to be relevant, with our kernel.

This Chapter is organized as follows: In the next Section we begin by defining so called "optimal assignment kernels" as a general class of symmetric similarity measures and show how they can be cast into proper positive semidefinite kernel functions. Given this result we can introduce our optimal assignment kernel for chemical graphs in Section 6.3 and show how it can be computed efficiently. In Sections 6.4 and 6.5 we investigate possible extensions of the optimal assignment kernel, namely by means of the reduced graph representation and by incorporating expert provided descriptor information. In Section 6.6 we show experimental results of our method compared to the marginalized graph kernel and to classical descriptor-based QSAR/QSPR models on the datasets from the last Chapter and show that in several cases we can significantly outperform marginalized graph kernels as well as descriptor-based models with and without performing automatic descriptor selection by means of RFE. Furthermore, we show that by combining our kernel with descriptors known to be relevant to the QSAR/QSPR problem at hand a further reduction of the prediction error is possible. We also demonstrate the good performance of the reduced graph representation. In Section 6.7 we briefly summarize and discuss our results, before in Section 6.8 we draw a general conclusion of our work and point out directions of future research.

6.2 Optimal Assignment Kernels

Let $\mathcal{X}$ be some domain of structured objects (e.g. graphs). Let us denote the parts of some object x (e.g. the nodes of a graph) by $x[1], ..., x[|x|]$, i.e. x consists of $|x|$ parts, while another object y consists of $|y|$ parts. Let $\mathcal{X}'$ denote the domain of all parts, i.e. $x[i] \in \mathcal{X}'$ for $1 \leq i \leq |x|$. Further let π be some permutation of either an $|x|$-subset of natural numbers $\{1, ..., |y|\}$ or an $|y|$-subset of $\{1, ..., |x|\}$ (this will be clear from context).

Definition 6.2.1. (Optimal Assignment Kernels)

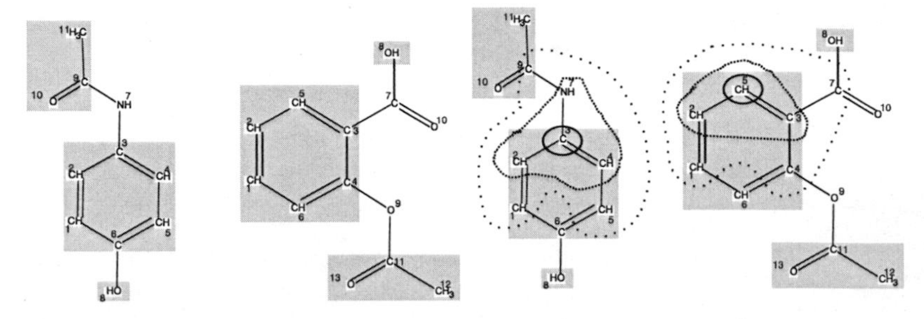

(a) Matching regions of two molecular graphs.

(b) Direct and indirect neighbors of atom 3 in the left and atom 5 in the right molecule.

Figure 6.1: Intuition of the optimal assignment kernel.

Let $k_1 : \mathcal{X}' \times \mathcal{X}' \rightarrow \mathbb{R}$ be some non-negative, symmetric and positive semidefinite kernel. Then $k_A : \mathcal{X} \times \mathcal{X} \rightarrow \mathbb{R}$ with

$$k_A(x, y) := \begin{cases} \max_\pi \sum_{i=1}^{|x|} k_1(x[i], y[\pi(i)]) & \text{if } |y| \geq |x| \\ \max_\pi \sum_{j=1}^{|y|} k_1(x[\pi(j)], y[j]) & \text{otherwise} \end{cases}$$

is called an *optimal assignment kernel.*

This definition captures the idea of a maximal weighted bipartite matching (optimal assignment) of the parts of two objects. Each part of the smaller of both structures is assigned to exactly one part of the other structure such that the overall similarity score between both structures is maximized (Fig. 6.2).

Lemma 6.2.2. *For all x: $k_A(x, x) = \sum_i k_1(x[i], x[i])$.*

Proof. Let $m = |x|$. For any π it is

$$\begin{aligned} & k_1(x[1], x[\pi(1)]) + \ldots + k_1(x[m], x[\pi(m)]) & (6.1) \\ \leq\ & \frac{1}{2}\left(k_1(x[1], x[1]) + k_1(x[\pi(1)], x[\pi(1)]) + \ldots + \right. & (6.2) \\ & \left. k_1(x[m], x[m]) + k_1(x[\pi(m)], x[\pi(m)])\right) & \\ =\ & \sum_i k_1(x[i], x[i]) & (6.3) \end{aligned}$$

because $2k_1(x[i], x[\pi(i)]) \leq k_1(x[i], x[i]) + k_1(x[\pi(i)], x[\pi(i)])$ for all i and because each term in (6.2) appears twice, once as $k_1(x[i], x[i])$ and once as $k_1(x[\pi(j)], x[\pi(j)])$ with

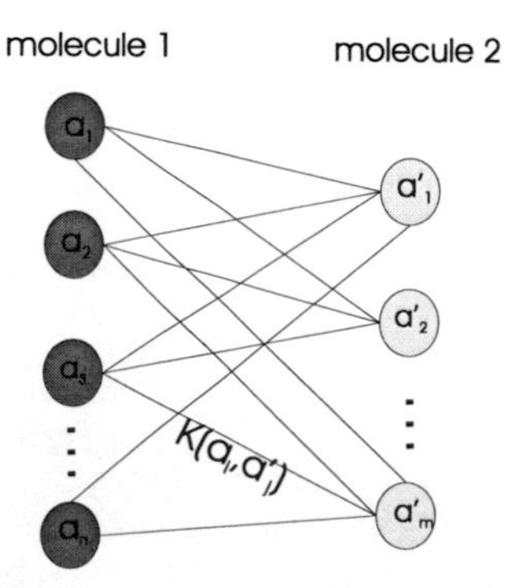

Figure 6.2: Possible assignments of atoms from molecule 2 to those of molecule 1. The kernel function k measures the similarity of a pair of atoms (a_i, a'_j) including information on structural and physico-chemical properties. The goal is to find the matching, which assigns each atom from molecule 2 to exactly one atom from molecule 1, such that the overall similarity score, i.e. the sum of edge weights in the bipartite graph, is maximized.

$\pi(j) = i$. This is a direct consequence of the positive semi-definiteness of k_1. If we now take the maximum over all π, then $(6.1) = k_A(x, x) = (6.2) = (6.3)$

□

Theorem 6.2.3. *k_A is a symmetric and positive semidefinite kernel for each pair of objects.*

Proof. Clearly, k_A is symmetric, because of the definition.

We now show that each 2×2 kernel matrix is positive semidefinite. W.l.o.g. let $x, y \in \mathcal{X}$ with $|y| \geq |x|$. Because of lemma 6.2.2, we have $k_A(x, x) = \sum_i k_1(x[i], x[i]),\ k_A(y, y) = \sum_j k_1(y[j], y[j])$. Further it holds for all $\alpha, \beta \in \mathbb{R}$ and i, j

$$2\alpha\beta k_1(x[i], y[j]) \leq \alpha^2 k_1(x[i], x[i]) + \beta^2 k_1(y[j], y[j]) \tag{6.4}$$

because k_1 is a positive semidefinite kernel. It is:

$$\begin{aligned} & \alpha^2 k_A(x, x) - 2\alpha\beta k_A(x, y) + \beta^2 k_A(y, y) & (6.5)\\ = \ & \alpha^2 \sum_i k_1(x[i], x[i]) - 2\alpha\beta \max_\pi \sum_i k_1(x[i], y[\pi(i)]) + \beta^2 \sum_j k_1(y[j], y[j]) \end{aligned}$$

By definition of k_A the second sum of (6.5) has $\min(|x|, |y|) = |x|$ addends. Let $y'[i]$ be the part of y to which $x[i]$ is assigned. Using the non-negativity of k_1 and (6.4) we have:

$$\alpha^2 \sum_i k_1(x[i], x[i]) - 2\alpha\beta \max_\pi \sum_i k_1(x[i], y[\pi(i)]) + \beta^2 \sum_j k_1(y[j], y[j]) \quad (6.6)$$

$$\geq \sum_{i=1}^{|x|} (\alpha^2 k_1(x[i], x[i]) - 2\alpha\beta k_1(x[i], y'[i]) + \beta^2 k_1(y'[i], y'[i])$$

$$\geq 0$$

□

Remark Theorem 6.2.3 proves the positive semi-definiteness of each 2×2 kernel matrix. However, this does not imply that also each $n \times n$ kernel matrix ($n > 2$) is positive semidefinite. As an example consider the matrix

$$\begin{pmatrix} 1 & 1 & 0 \\ 1 & 1 & 1 \\ 0 & 1 & 1 \end{pmatrix}$$

which is **not** positive semidefinite, since it has determinant -1. Nevertheless all 2×2 submatrices are positive semidefinite.

As it is highly desirable to have a positive semidefinite kernel matrix for SVM training (c.f. Chapter 2.5), a way out is given by shifting the spectrum of the $n \times n$ kernel matrix on the training data. This can be achieved by transforming the original kernel matrix $\mathbf{K}$ as

$$\mathbf{K} \leftarrow \mathbf{K} - \lambda_{min}\mathbf{I} \quad (6.7)$$

where λ_{min} is the smallest negative eigenvalue of $\mathbf{K}$, if there is any (and zero otherwise). To see this consider the eigenvector decomposition of $\mathbf{K}$:

$$\mathbf{K} - \lambda_{min}\mathbf{I} = \mathbf{V}\mathbf{\Lambda}\mathbf{V}^T - \lambda_{min}\mathbf{I} = \mathbf{V}\mathbf{\Lambda}\mathbf{V}^T - \lambda_{min}\mathbf{V}\mathbf{V}^T = \mathbf{V}(\mathbf{\Lambda} - \lambda_{min}\mathbf{I})\mathbf{V}^T \quad (6.8)$$

where $\mathbf{V}$ is an orthogonal matrix with columns being the eigenvectors and Λ a diagonal matrix containing the eigenvalues.

In the following we will always assume that for the optimal assignment kernel, when applied to a set of objects, transformation (Eq.: 6.7) has been performed.

6.3 Optimal Assignment Kernels for Chemical Compounds

We now turn to the construction of an optimal assignment kernel for molecular graphs. Let us assume we have two molecules M and M', which have atoms $a_1, ..., a_n$ and $a'_1, ..., a'_m$. Let us further assume we have some non-negative kernel k_{nei}, which compares a pair of atoms $(a_h, a'_{h'})$ from both molecules, including information on their neighborhoods, membership to certain substructures (like aromatic systems, donors, acceptors, and so on) and other physico-chemical properties (e.g. mass, partial charge and others). Then the similarity between M and M' can be calculated using the optimal assignment kernel

$$k_A(M, M') = \begin{cases} \max_\pi \sum_{h=1}^{n} k_{nei}(a_h, a'_{\pi(h)}) & \text{if } m \geq n \\ \max_\pi \sum_{h'=1}^{m} k_{nei}(a_{\pi(h')}, a'_{h'}) & \text{otherwise} \end{cases} \tag{6.9}$$

That means we assign each atom of the smaller of both molecules to exactly one atom of the bigger molecule such that the overall similarity score is maximized. The optimal assignment is a classical problem from graph theory and can be computed efficiently in $\mathcal{O}(\max(n, m)^3)$ (Mehlhorn and Näher, 1999). Although this seems to be a drawback compared to the quadratic time complexity of marginalized graph kernels, we have to point out, that marginalized graph kernels have to be iteratively computed until convergence, and thus in practice, depending on the size of the molecules, there might be no real difference in computation time.

In order to prevent larger molecules automatically to achieve a higher kernel value than smaller ones, we should further normalize our kernel (Schölkopf and Smola, 2002), i.e.

$$k_A(M, M') \leftarrow \frac{k_A(M, M')}{\sqrt{k_A(M, M) k_A(M', M')}} \tag{6.10}$$

Finally, given the results from the last Section, before SVM training we should transform the kernel matrix on our training data to assure its positive semi-definiteness. (For testing no transformation is necessary.)

We now have to define the kernel k_{nei}. For this purpose let us suppose we have non-negative kernels k_{atom} and k_{bond}, which compare the atom and bond labels, respectively. A natural choice for both kernels would be the RBF-kernel. The set of labels associated with each atom or bond can be interpreted as feature vectors. As the individual features for an atom or bond can live on different numerical scales, it is beneficial to normalize them, e.g. to mean 0 and standard deviation 1. Let us denote by $a \to n_i(a)$ the bond connecting atom a with its ith neighbor $n_i(a)$. Let us further denote by $|a|$ the number of neighbors of atom a. We now define a kernel R_0, which compares all direct neighbors of atoms (a, a') as the optimal assignment kernel between all neighbors of a and a' and the bonds leading to them, i.e.

$$R_0(a, a') = \frac{1}{|a|} \max_\pi \sum_{i=1}^{|a|} k_{atom}(n_i(a), n_{\pi(i)}(a')) \cdot k_{bond}(a \to n_i(a), a' \to n_{\pi(i)}(a')) \tag{6.11}$$

where we assumed $|a| \geq |a'|$ for the sake of simplicity of notation. As an example consider the C-atom 3 in the left and the C-atom 5 in the right structure of Fig. 6.1: If our only atom and bond features were the element type and bond order, respectively, and k_{atom} and k_{bond} would simply count a match by 1 and a mismatch by 0, our kernel $R_0(a_3, a'_5)$ would tell us that 2 of 3 possible neighbors of atom 3 in the left structure match with the neighbors of atom 5 in the right structure. It is worth mentioning that the computation of R_0 can be done in $\mathcal{O}(1)$ time complexity as for chemical compounds $|a|$ and $|a'|$ can be upper bounded by a small constant (e.g. 4).

Of course it would be beneficial not to consider the match of direct neighbors only, but also that of indirect neighbors and atoms having a larger topological distance. For this purpose we can evaluate R_0 not at (a, a') only, but also at all pairs of neighbors, indirect neighbors and so on, up to some topological distance L. In our example that would mean we also evaluate $R_0(a_2, a'_2), R_0(a_4, a'_2), R_0(a_7, a'_2), \ldots$ and so on. If we consider the mean of all these values and add them to $k_{atom}(a, a') + R_0(a, a')$, this leads to the following definition of the kernel k_{nei}:

$$k_{nei}(a, a') = k_{atom}(a, a') + R_0(a, a') + \sum_{\ell=1}^{L} \gamma(\ell) R_\ell(a, a') \tag{6.12}$$

Here R_ℓ denotes the mean of all R_0 evaluated at neighbors of topological distance ℓ, and $\gamma(\ell)$ is a decay parameter, which reduces the influence of neighbors that are further away and depends on the topological distance ℓ to (a, a'). It makes sense to set $\gamma(\ell) = p(\ell)p'(\ell)$, where $p(\ell), p'(\ell)$ are the probabilities for molecules M, M' that neighbors with topological distance ℓ are considered.

A key observation is that R_ℓ can be computed efficiently from $R_{\ell-1}$ via the recursive relationship

$$R_\ell(a, a') = \frac{1}{|a||a'|} \sum_{i,j} R_{\ell-1}(n_i(a), n_j(a')) \tag{6.13}$$

I.e. we can compute k_{nei} by iteratively revisiting all direct neighbors of (a, a') only. In case that L is set to a constant we thus still have a $\mathcal{O}(1)$ time complexity for the calculation of k_{nei}. In case that $L \to \infty$, we can prove the following theorem:

Theorem 6.3.1. *Let $\gamma(\ell) = (\hat{p}_1\hat{p}_2)^\ell$ with $\hat{p}_1, \hat{p}_2 \in (0, 1)$. If there exists a $C \in \mathbb{R}^+$, such that $k_{atom}(a, a') \leq C$ for all a, a' and $k_{bond}(a \to n_i(a), a' \to n_j(a')) \leq C$ for all $a \to n_i(a), a' \to n_j(a')$, then (6.12) converges for $L \to \infty$.*

Proof. It is

$$R_0(a, a') \leq \frac{\min(|a|, |a'|)}{\max(|a|, |a'|)} C^2 \leq C^2 \tag{6.14}$$

and thus

$$R_1(a, a') \leq \frac{1}{|a||a'|} \sum_{i=1}^{|a|} \sum_{j=1}^{|a'|} C^2 = C^2 \tag{6.15}$$

Hence, also $R_\ell(a, a') \leq C^2$ for $\ell = 2, ..., L$. Therefore we have

$$(6.12) \leq C + C^2 + (\hat{p}_1\hat{p}_2)^1 C^2 + ... + (\hat{p}_1\hat{p}_2)^L C^2 = C + C^2 + C^2 \sum_{\ell=1}^{L} (\hat{p}_1\hat{p}_2)^\ell \quad (6.16)$$

which converges for $L \rightarrow \infty$.

□

Note, that the boundedness of k_{atom} and k_{bond} is especially fulfilled, if we take the RBF-kernel for both.

To briefly summarize, our method works as follows: We first compute the similarity of all atom and bond features using the kernels k_{atom} and k_{bond}. Having these results, we can compute the match of direct neighbors R_ℓ for each pair of atoms from both molecules by means of (6.11). From R_0 we can compute $R_1, ..., R_L$ by iteratively revisiting all direct neighbors of each pair of atoms and computing the recursive update formula (6.13). Having k_{atom} and $R_1, ..., R_L$ directly gives us k_{nei}, the final similarity score for each pair of atoms, which includes information on memberships to substructures as well as physico-chemical properties. With k_{nei} we finally compute the optimal assignment kernel between two molecules M and M' using (6.9) and (6.10). Thereby, in our present implementation (Eq.: 6.9) is calculated by the *Hungarian method* (Kuhn, 1955). Alternatively one can use the algorithm described in Mehlhorn and Näher (1999). We preferred the Hungarian method simply because it is more commonly used for solving the optimal assignment problem. In a last step, before SVM training, the kernel matrix is made positive semidefinite, as explained in Section 6.2.

6.4 Incorporation of Relevant Descriptor Information

6.4.1 Considered Approaches

For some QSAR/QSPR problems it is known that certain molecular descriptors are crucial (c.f. Chapter 5). E.g. for the human intestinal absorption the polar surface area of molecules plays an important role (Artursson and Bergström, 2003; van de Waterbeemd and Gifford, 2003). In some sense this descriptor describes global properties of a molecule, whereas our kernel relies on the graph structure and hence on local properties of a molecule. In this way our kernel describes the similarity of two molecules in a different way than it is done by comparing global molecular descriptors. It seems obvious that if a-priori knowledge on certain relevant descriptors is available, then it should be incorporated into the QSAR/QSPR model. Such a combination would allow us to integrate different notions of similarity between two molecules into one model.

A straight-forward way of doing so is to consider the sum of a RBF-kernel for the descriptor information we have from our expert knowledge and the optimal assignment kernel.

Another way would be to consider multi-kernel SVMs (Weston, 1999). The idea here is to learn a decision function of the form

$$f(x) = \operatorname{sgn}\left(\sum_{i=1}^{n}\sum_{r=1}^{\kappa}\alpha_{ir}k_r(x_i,x)+b\right) \tag{6.17}$$

where α_{ir} is the Lagrangian multiplier associated to the ith training pattern with the rth kernel. That means the input patterns are now mapped not into one, but into κ different feature spaces in parallel, and all constructed hyperplanes in them are combined to give the final decision surface. For SVMs the consequence is that the margin has to be maximized in all κ feature spaces. This approach can be used for classification as well as for regression estimation. Our current implementation of the multi-kernel SVM uses a linear programming formulation, which in parallel determines all Lagrangian multipliers α_{ir} using the MATLAB^{TM} "linprog" solver. While this is clearly not optimal from the computational point of view, it should nonetheless enable us to study the principle usefulness of the multi-kernel SVM approach on at least the smaller of our datasets.

A different approach was presented in Lanckriet et al. (2004): The idea here is to learn the kernel matrix as a linear combination $\mathbf{K} = \sum_{r=1}^{\kappa}\mu_r\mathbf{K}_r$ of κ base kernel matrices, i.e. we have a weight μ_r for each kernel. The optimization of the weight vector μ is performed together with that of the Lagrangian multipliers α in the SVM decision function via semidefinite programming[1] (Vandenberghe and Boyd, 1996). The approach is originally developed only for the classification case, but as our QSAR/QSPR problems are often regression problems, we derive a version for Kernel Ridge Regression (c.f. Chapter 3.3) in Subsection 6.4.2. Semidefinite programs can be solved using general purpose programs such as SeDuMi (Sturm, 1999) in $\mathcal{O}((\kappa+n)^2n^{2.5})$ time. To bring the (in-)equalities in the required form, one can use YALMIP (Löfberg, 2004). Like for multi-kernel SVMs, SeDuMi as an off-the-shelf solver enables us to study the principle usefulness of the semidefinite programming approach, but is clearly not suitable for large scale applications.

Finally, if we assume that the feature spaces induced by the optimal assignment kernel and by the relevant descriptor information via a RBF kernel share some common structure, then one can also employ KCCA to extract this information. The use of KCCA is further motivated by several successful real world applications with heterogeneous data representations in the literature (e.g. Yamanishi et al., 2004a,b).

[1] Semidefinite programming deals with optimization problems over symmetric positive definite matrices with linear cost function and linear constraints. Linear programming and quadratic programming can be understood as special cases of semidefinite programming.

6.4.2 Learning the Kernel Matrix in Kernel Ridge Regression via Semidefinite Programming

Let us now turn to the question, how the kernel matrix can be learned as a linear combination of base kernels via semidefinite programming in Kernel Ridge Regression. Thereby, we follow an argumentation similar to Lanckriet et al. (2004):

Let $\succeq 0$ denote the positive semi-definiteness of a matrix. Let $\mathcal{K}$ denote the set of positive semidefinite kernel matrices with constant trace c that can be expressed as a linear combination of kernel matrices from the set $\{\mathbf{K}_1, ..., \mathbf{K}_\kappa\}$. That is, $\mathcal{K}$ is the set of matrices $\mathbf{K}$ satisfying

$$\begin{aligned} \mathbf{K} &= \sum_{r=1}^{\kappa} \mu_r \mathbf{K}_r \qquad (6.18) \\ \mathbf{K} &\succeq 0 \\ \text{trace}(\mathbf{K}) &= c \end{aligned}$$

As already mentioned in Chapter 3.3, the optimal Lagrangian multipliers α^* of the Kernel Ridge Regression problem are obtained by the solution of the optimization problem

$$\min_{\alpha} -\alpha^T(\mathbf{K} + \tau\mathbf{I})\alpha + 2\alpha^T\mathbf{y} \qquad (6.19)$$

yielding $\alpha^* = (\mathbf{K} + \tau\mathbf{I})^{-1}\mathbf{y}$. We are now interested in the matrix $\mathbf{K} \in \mathcal{K}$ that minimizes (6.19), i.e.

$$\min_{\mathbf{K}\in\mathcal{K}} \min_{\alpha} -\alpha^T(\mathbf{K} + \tau\mathbf{I})\alpha + 2\alpha^T\mathbf{y} \qquad (6.20)$$

Obviously, this is a convex optimization problem in $\mathcal{K}$ and can thus be cast into a semidefinite program.

Theorem 6.4.1. *The kernel matrix $\mathbf{K} \in \mathcal{K}$ that optimizes (6.19) with $\tau \geq 0$ can be found by solving the following semidefinite program:*

$$\begin{aligned} \min_{\mu} \quad & t \qquad (6.21) \\ \text{s.t.} \quad & \sum_{r=1}^{\kappa} \mu_r \mathbf{K}_r \succeq 0 \\ & \textit{trace}\left(\sum_{r=1}^{\kappa} \mu_r \mathbf{K}_r + \tau\mathbf{I}\right) = c \\ & \begin{pmatrix} \sum_{r=1}^{\kappa} \mu_r \mathbf{K}_r + \tau\mathbf{I} & \mathbf{y} \\ \mathbf{y}^T & t \end{pmatrix} \succeq 0 \end{aligned}$$

Proof. Putting the solution $\alpha^* = (\mathbf{K} + \tau\mathbf{I})^{-1}\mathbf{y}$ back into (6.19) leads to the optimum $\mathbf{y}^T(\mathbf{K} + \tau\mathbf{I})^{-1}\mathbf{y}$. We are now interested in

$$\min_{\mathbf{K}\in\mathcal{K}} \quad \mathbf{y}^T(\mathbf{K} + \tau\mathbf{I})^{-1}\mathbf{y} \tag{6.22}$$

$$\Leftrightarrow \min_{\mu} \quad \mathbf{y}^T\left(\sum_{r=1}^{\kappa}\mu_r\mathbf{K}_r + \tau\mathbf{I}\right)^{-1}\mathbf{y} \tag{6.23}$$

$$\text{s.t.} \quad \sum_{r=1}^{\kappa}\mu_r\mathbf{K}_r \succeq 0$$

$$\text{trace}\left(\sum_{r=1}^{\kappa}\mu_r\mathbf{K}_r + \tau\mathbf{I}\right) = c$$

which is a convex optimization problem. We can thus rewrite the optimization problem as:

$$\min_{\mu,t} \quad t \tag{6.24}$$

$$\text{s.t.} \quad t \geq \mathbf{y}^T\left(\sum_{r=1}^{\kappa}\mu_r\mathbf{K}_r + \tau\mathbf{I}\right)^{-1}\mathbf{y}$$

$$\sum_{r=1}^{\kappa}\mu_r\mathbf{K}_r \succeq 0$$

$$\text{trace}\left(\sum_{r=1}^{\kappa}\mu_r\mathbf{K}_r + \tau\mathbf{I}\right) = c$$

Using the Schur Complement Lemma[2] the condition $\mathbf{y}^T\left(\sum_{r=1}^{\kappa}\mu_r\mathbf{K}_r + \tau\mathbf{I}\right)^{-1}\mathbf{y} \leq t$ can equivalently be expressed as

$$\begin{pmatrix} \sum_{r=1}^{\kappa}\mu_r\mathbf{K}_r + \tau\mathbf{I} & \mathbf{y} \\ \mathbf{y}^T & t \end{pmatrix} \succeq 0 \tag{6.25}$$

which proves the theorem.

□

6.5 Reduced Graph Representation

The main intuition of our method lies in the matching of substructures from both molecules. In Section 6.3 we achieved this by using structural, neighborhood and other characteristic

[2]**Schur Complement Lemma**: *Let* $\mathbf{A}, \mathbf{C}$ *be square and symmetric matrices. If* $\mathbf{A}$ *is positive definite, then* $\mathbf{X} = \mathbf{X}^T = \begin{pmatrix} \mathbf{A} & \mathbf{B} \\ \mathbf{B}^T & \mathbf{C} \end{pmatrix} \succeq 0$ *if and only if* $\mathbf{S} = \mathbf{C} - \mathbf{B}^T\mathbf{A}^{-1}\mathbf{B} \succeq 0$.

information for each single atom and bond, and computing the optimal assignment kernel between atoms of both molecules. A natural extension of this idea is to collapse structural features, like rings, donors, acceptors and others, into a single node of the graph representation of a molecule. Atoms not matching a-priori defined types of structural features can even be removed (Martin, 1998). This allows us to concentrate on important structural elements of a molecule, where the definition of what an important structural element actually is, depends on the QSAR/QSPR problem at hand and could be given by expert knowledge e.g. in form of so-called SMARTS[3,4] patterns. The high relevance of such a pharmacophore mapping for QSAR/QSPR models is also reported in Chen et al. (1999) and Oprea et al. (2002). If atoms match more than one SMARTS pattern, a structural feature consists of the smallest substructure that cannot be further divided into subgroups with regard to all patterns. That means in our reduced graph we may get a substructure node describing a ring only and another one describing both, a ring and an acceptor.

Two principle problems have to be solved to implement the reduced graph: First, if certain atoms are removed from the molecular graph, then we may obtain nodes, which are disconnected from the rest of the graph. They have to be reconnected by new edges again such that these new edges preserve the neighborhood information, i.e. if before we had $a \rightarrow b$ and $b \rightarrow c$ and atom b is removed, we should obtain $a \rightarrow c$. These new edges should contain information on the topological and geometrical distance of the substructures connected by them. Thereby the topological distance between two substructures is calculated as the minimal topological distance between the atoms belonging to them, whereas the geometrical distance is computed between the centers of gravity in order to conserve information on the 3D structure of the substructures (Fig. 6.3).

Second, we have to define how the feature vectors for each single atom and bond included in a substructure can be transferred to the whole substructure. This can, for instance, be solved by recursively applying our method from the last Section, if two substructures have to be compared.

A principle advantage of the reduced graph representation lies in the fact that complete substructures and their neighbor substructures can be compared at once. This allows us to concentrate on relevant structural aspects of the molecular graph. By means of SMARTS patterns in principle it is possible to define arbitrary structural features to be condensed in one node of the reduced molecular graph. That means in some sense one can change the "resolution" at which one looks at the molecule. This way one achieves an even higher flexibility as offered by feature trees, because rather than considering the average over atom and bond features contained in a substructure, substructure nodes are compared on an atomic level and hence less structural information is lost. From a computational side a reduced graph representation may be useful, because the effort for computing the optimal assignment is reduced.

[3]Daylight Chemical Information Systems Inc., http://www.daylight.com

[4]SMARTS, similar to regular expressions, is a language for describing patterns on molecular graphs.

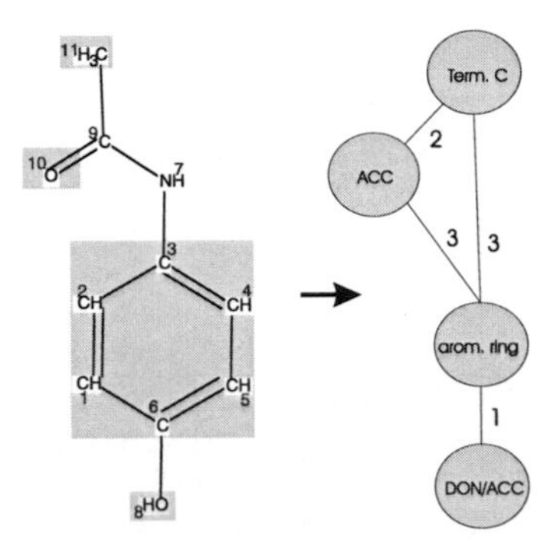

Figure 6.3: Example of a conversion of a molecule into its reduced graph representation with edge labels containing the topological distances.

6.6 Experiments

6.6.1 Datasets

We used the same five datasets as introduced in the last Chapter. For the calculation of the optimal assignment kernel we used the same atom properties that were used for the descriptor calculations (Table E.1 in the Appendix).

6.6.2 Results

Example Matching of Two Molecules Before turning to the evaluation results, in Fig. 6.4 we show an optimal assignment calculated by our method for the two example molecules, which were taken from the HIA dataset. As one can see, the optimal assignment indeed nicely matches the ring atoms and the atoms of the carbonyl groups and thus implements the intuition explained in the introduction.

Smallest Eigenvalues According to our results from Section 6.2 the optimal assignment kernel from Section 6.2 for a set of n molecules without spectrum shifting cannot expected to be positive semidefinite. In Fig. 6.5 we depict the smallest eigenvalues calculated for the optimal assignment kernel (OA) matrix before spectrum shifting on our five datasets during a 10-fold cross-validation procedure. The same was done for the optimal assignment kernel using the reduced graph representation (OARG). As seen from Fig. 6.5 the smallest eigenvalues are always relatively small in magnitude, i.e. the optimal assignment kernel is "almost" positive semidefinite. This may be due to the fact that all 2×2 submatrices of the full $n \times n$ kernel matrix are positive semidefinite. Practically this

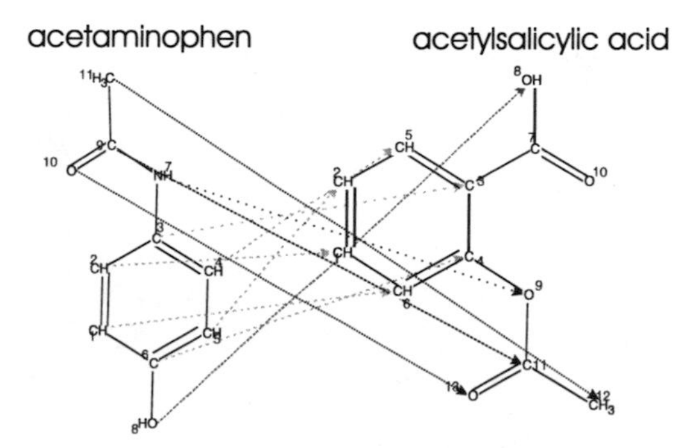

Figure 6.4: Two molecules from the HIA dataset and the optimal assignment computed by our method.

implies that the spectrum shifting has only a minor effect. That means we are luckily not in the situation, where after spectrum shifting the diagonal of the kernel matrix exceeds the rest of the matrix entries by magnitudes, which would cause an "over-regularization" or underfitting problem (comparable to Kernel Ridge Regression with regularization constant τ being very large).

Comparison to Descriptor-Based Models and the Marginalized Graph Kernel Let us now turn to the evaluation of our method. We compared the optimal assignment kernel to the marginalized graph kernel (MG) using the same atom and bond features, to the full descriptor model (DESC) and the standard RFE based descriptor selection model from the last Chapter (DESCSEL). For the OA and the MG kernel we considered the kernels k_{atom} and k_{bond} as a product of two RBF kernels for the real valued and the nominal atom/bond features. The width of these RBF kernels was set such that $\exp(-D/(2\sigma^2)) = 0.1$ (D = number of atom/bond features). Thereby all atom and bond features were approximately normalized to mean 0 and standard deviation 1 by considering the means and standard deviations of the atom and bond properties over all our datasets. Furthermore, we explicitly set k_{bond} to 0, if one bond was in an aromatic system and the other not, or if both bonds had a different bond order. This corresponds to the multiplication with a δ-kernel. The probabilities to reach neighbors with topological distance ℓ was set to $p(\ell) = p'(\ell) = 1 - \frac{1}{L}$ with $L = 3$. This allows us to consider the neighborhood of a whole 6-ring of an atom. We used a SVM on the HIA, Yoshida and SOL classification datasets and a Support Vector Regression (ϵ−SVR) on the BBB and Huuskonen regression problems. The prediction strength was evaluated by means of 10-fold cross-validation. Thereby on the classification problems we ensured that the ratio of examples from both classes in the actual

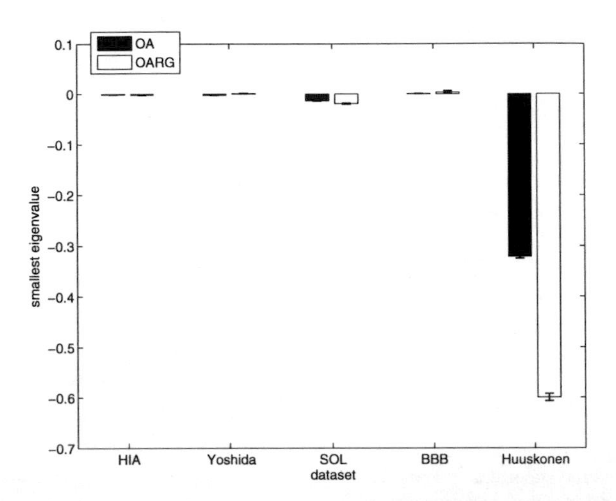

Figure 6.5: Smallest eigenvalues of the kernel matrix calculated on the training data during a 10-fold cross-validation procedure.

training set was always the same (stratified cross-validation). On each actual training fold a model selection for the necessary parameters was performed by evaluating each candidate parameter set by an extra level of 10-fold cross-validation. For the MG kernel the model selection included testing termination probabilities $p_t = 0.1, 0.3, 0.5, 0.7$. The soft-margin parameter C was chosen from the interval $[2^{-2}, 2^{14}]$. On the regression datasets (BBB, Huuskonen) the width of the ϵ-tube was chosen from the interval $[2^{-8}, 2^{-1}]$. All descriptor values (also the target values logBB, logS in case of the regression datasets) were normalized to mean 0 and standard deviation 1 on each training fold, and the calculated scaling parameters were then applied to normalize the descriptor values in the actual test set.

Fig. 6.6 - 6.12 and Table E.2 in the Appendix show the results we obtained. Using our OA kernel we outperformed the DESC model statistically significant on the SOL, the BBB and the Huuskonen dataset ($p < 0.1\%$ for MSE and r^2). Thereby statistical significance was tested by a two-tailed paired t-test at significance level 10%. On the BBB dataset we also achieved significantly better results than the DESCSEL model ($p = 6.49\%$ for MSE, $p = 3.56\%$ for r^2). At the same time our OA kernel circumvented the high computational burden of computing thousands of descriptors followed by a descriptor selection algorithm.

Compared to the MG kernel our OA kernel utilizing the *Hungarian method* for the assignment calculation was significantly better on the Yoshida dataset ($p = 6.57\%$), the

Table 6.1: Average computation times (ms $\pm$ standard deviation) for one kernel function evaluation for the different kernels.

Method	HIA	Yoshida	SOL	BBB	Huuskonen
MG	5.66 ± 6.34	4.4 ± 3.74	6.64 ± 4.86	3.14 ± 10.12	1.87 ± 1.81
OA	$\mathbf{4.90 \pm 7.22}$	$\mathbf{3.33 \pm 3.63}$	$\mathbf{5.04 \pm 7.09}$	$\mathbf{2.81 \pm 9.50}$	$\mathbf{1.58 \pm 1.47}$
OARG	5.92 ± 9.00	4.23 ± 8.30	5.61 ± 5.07	3.38 ± 13.68	1.83 ± 1.86

BBB dataset (p = 4.65% according to MSE) and the Huuskonen dataset ($p < 1\%$ for MSE/r^2). At the same time using our JAVA implementation on an AMD single core Opteron 64 Bit 2.4GHz machine the computation time for the OA kernel was always **below** that for the MG kernel (Table 6.1).

Reduced Graph Representation We also investigated the effect of the reduced graph representation (OARG kernel — Fig. 6.6 - 6.12 and last row in Table E.2). Thereby in the reduced graph representation only direct neighbors were considered to compute (i.e. $L' = 1$), whereas for the comparison of nodes representing structural elements we used $L = 3$ atomic neighbors as before. We considered the following pharmacophore features defined by SMARTS patterns: ring, donor, donor/acceptor, terminal carbon, positive charge, negative charge. The precise definitions of these patterns can be found in Table E.6 in the Appendix. Molecules, which did not contain any of these features and hence lead to an empty graph, were represented by the pattern "*", i.e. as one node. This only effected two compounds in the BBB dataset: N_2 and C_2HF_3BrCl.

Except for the Huuskonen dataset this choice of pharmacophoric features lead to similar cross-validation results of the OARG kernel than the original OA kernel. Obviously, on the Huuskonen dataset the defined SMARTS patterns did not capture the underlying chemistry in the right way. However, the improvement to the MG kernel was significant on the HIA dataset (p = 9.79%), the Yoshida dataset (p = 3.19%) and the BBB dataset (p = 0.71% for MSE). Likewise, the OARG kernel significantly outperformed the DESC and the DESCSEL model on the BBB dataset ($p < 0.1\%$ for MSE/r^2, p = 7.11% for MSE) and the DESC model on the SOL dataset ($p < 0.1\%$). This demonstrates that on these datasets the reduced graph representation, although using less structural information than the original OA kernel, covers well the relevant chemical and biological aspects of the molecules.

The computation times for the OARG kernel are slightly higher than those for the OA kernel, because of the effort needed to match the SMARTS patterns and to construct the reduced graph, but the differences are not significant (Table 6.1).

Comparison to Literature and Non-Kernel Methods We already discussed the problem to compare our results to those from the literature in the last Chapter (see Table

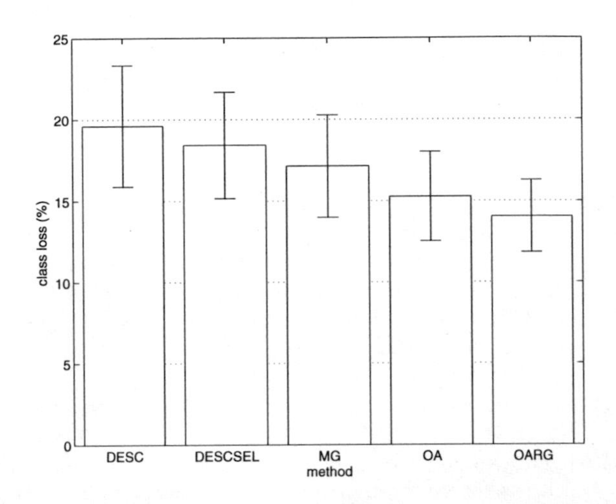

Figure 6.6: OA/OARG-Kernel vs. marginalized graph kernel and descriptor-based models: 10-fold cross-validation results (class loss in %) for the HIA dataset.

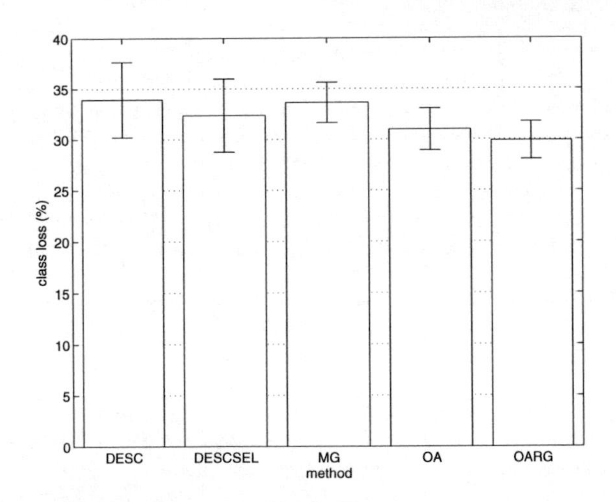

Figure 6.7: OA/OARG-Kernel vs. marginalized graph kernel and descriptor-based models: 10-fold cross-validation results (class loss in %) for the Yoshida dataset.

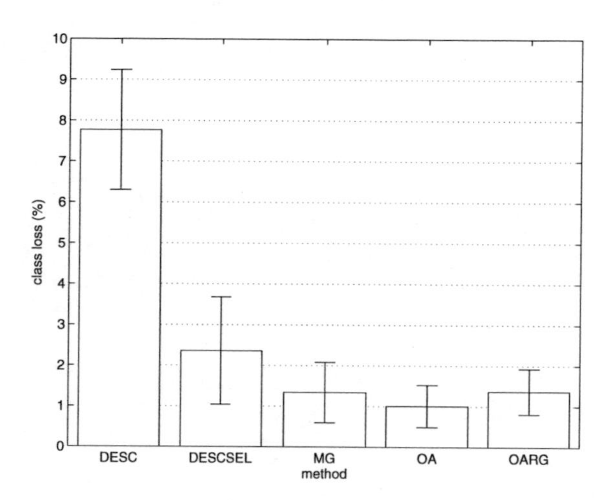

Figure 6.8: OA/OARG-Kernel vs. marginalized graph kernel and descriptor-based models: 10-fold cross-validation results (class loss in %) for the SOL dataset.

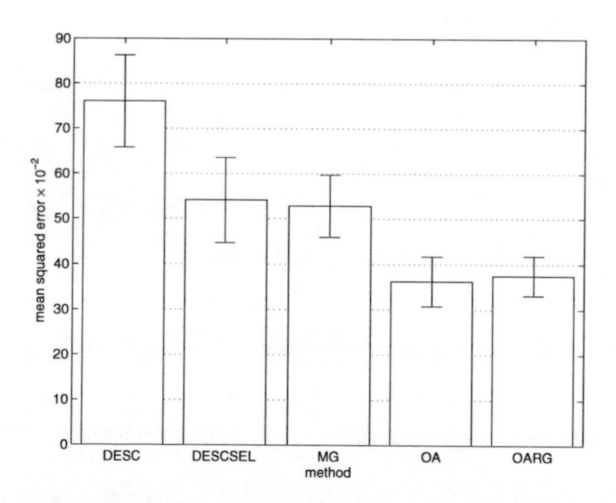

Figure 6.9: OA/OARG-Kernel vs. marginalized graph kernel and descriptor-based models: 10-fold cross-validation results (mean squared error) for the BBB dataset.

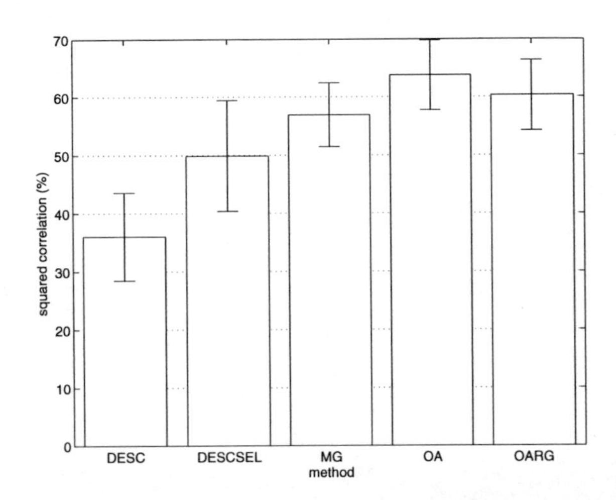

Figure 6.10: OA/OARG-Kernel vs. marginalized graph kernel and descriptor-based models: 10-fold cross-validation results (squared correlation in %) for the BBB dataset.

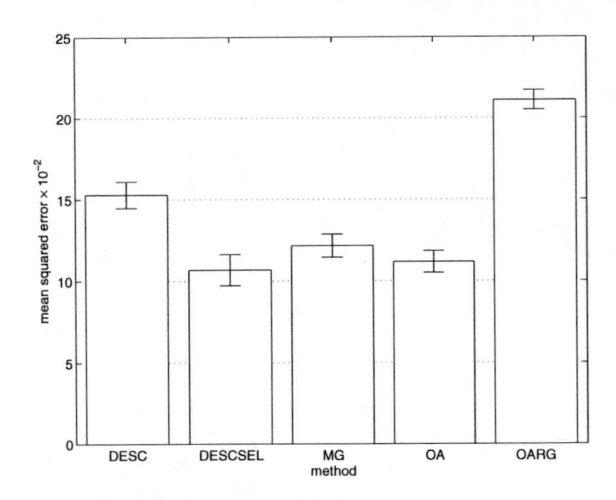

Figure 6.11: OA/OARG-Kernel vs. marginalized graph kernel and descriptor-based models: 10-fold cross-validation results (mean squared error) for the Huuskonen dataset.

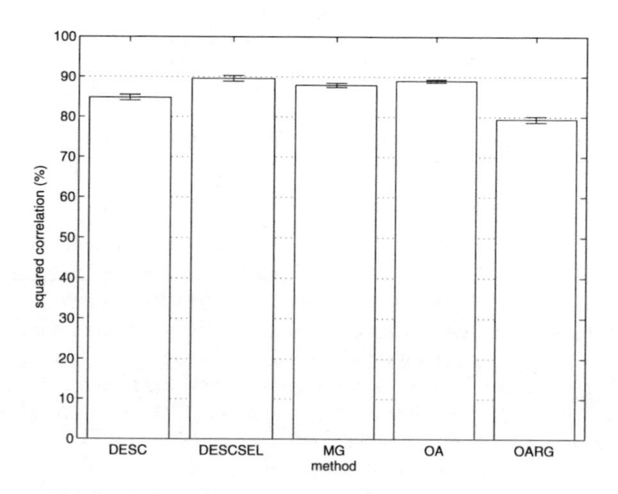

Figure 6.12: OA/OARG-Kernel vs. marginalized graph kernel and descriptor-based models: 10-fold cross-validation results (squared correlation in %) for the Huuskonen dataset.

5.2). If we ignore this issue, then our results on the BBB and the Huuskonen dataset are roughly in the same range as those reported in Hou and Xu (2003) and Huuskonen (2000). This should, however, taken with great caution for the reasons mentioned in 5.4.2. A better comparison is possible to the CART tree model in Chapter 5: The CART tree performs significantly worse than the OA and OARG kernels on the Yoshida ($p = 2.56\%, p = 2.93\%$), the SOL (both $p \leq 0.1\%$), the BBB ($p = 2.26\%, p = 3.36\%$ for MSE) and the Huuskonen dataset (both $p < 0.1\%$). These results thus confirm our findings from the last Chapter, where kernel-based methods were found to be superior to the CART tree as well.

Incorporation of Expert Knowledge Next we investigated the effect of incorporating expert knowledge into our approach. Using the experimental framework as described above we combined the OA/OARG kernel with the RBF kernel resulting from that model from the last Chapter, which consists only of descriptors known to be relevant for the specific dataset (EXP model). This was done by taking either the sum in combination with an ordinary SVM (OA/OARG+EXP), using the multi-kernel SVM (OA/OARG+MKSVM), the semidefinite programming approach (OA/OARG+SDP) or KCCA (OA/OARG+KCCA). Thereby we just fixed σ according to the heuristic described above.

Concerning KCCA, the regularization constant τ was fixed to 100 and the number N of canonical correlates chosen such at least 95% of the total covariance could be ex-

Table 6.2: Computation times (s) for one training using the semidefinite programming (SDP) and the multi-kernel SVM (MKSVM) approach with our current implementation. In all cases the parameters $C = 1$ and, in case of the OA+MKSVM regression, $\epsilon = 0.1$ were used.

Method	HIA	Yoshida	BBB	Huuskonen
OA+MKSVM	1.6	5.9	4.5	4354.2
OA+SDP	10.4	24.7	1.9	6481.2

plained. The extracted KCCA features from both representations were concatenated to form one vector space of dimension $2N$ and a SVM/SVR with RBF kernel was then trained on this representation. Thereby model selection and feature normalization was performed as described in the last Chapter for the DESC and DESCSEL methods.

Fig. 6.13 - 6.18 and Table E.5 in the Appendix show the results of our experiments. Due to the high computation time for the multi-kernel SVM using MATLAB's™ "linprog" solver and SeDuMi for the semidefinite programming approach, we are not showing the experimental outcomes of the OA/OARG+SDP and the OA/OARG+MKSVM method on the Huuskonen dataset (c.f. Table 6.2). This issue was already discussed in Subsection 6.4.2. Besides this fact, except for the KCCA technique incorporating expert knowledge in most cases lead to a reduction of the error rate. The difference to the original OARG kernel was statistically significant on the BBB dataset for the multi-kernel SVM (p = 5.45% for MSE, $p = 5.92\%$ for r^2) and the semidefinite programming approach ($p = 6.24\%$ for MSE). It was also significant on the Huuskonen dataset ($p < 0.1\%$ for MSE/r^2). Furthermore, in contrast to the original OA/OARG kernels, OA+EXP and OARG+EXP lead to a significant improvement compared the full descriptor model on the HIA dataset (p = 7.9%, 8.81%). Compared to the EXP model the improvement was significant on the Yoshida dataset for the OARG+MKSVM, the OA+SDP and the OARG+SDP model (p = 5.25%, 1.03%, 0.88%), and on the Huuskonen dataset for both models ($p < 0.1\%$ for MSE/r^2). In contrast to these positive results, OA+KCCA lead to a significant loss of prediction performance on the Huuskonen dataset ($p < 0.1\%$ for MSE/r^2).

6.7 Discussion

The above results revealed that a SVM trained with the optimal assignment kernel without any time intensive tuning of kernel parameters is able to achieve results, which are at least as good as those obtained with a descriptor-based SVM model with state-of-the-art descriptor selection. Our results are also significantly better than those obtained with a CART tree in the last Chapter. Comparisons to the marginalized graph kernel in several cases showed a significant improvement by our method. At the same time computation

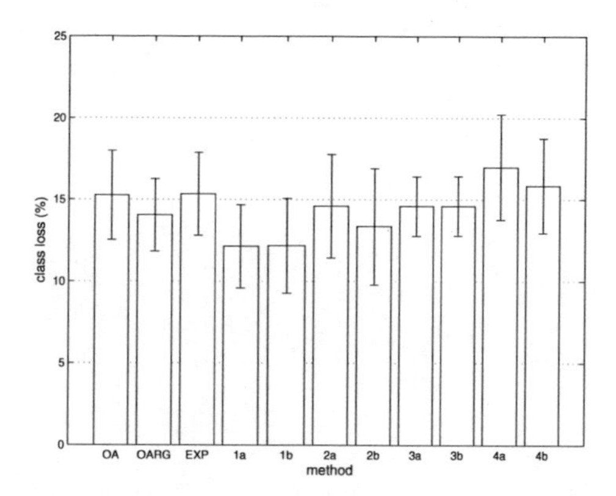

Figure 6.13: Effect of incorporating expert knowledge: 10-fold cross-validation results (class loss in %) for the HIA dataset. 1a/b = sum of OA/OARG with EXP model (OA/OARG+EXP); 2a/b = multi-kernel SVM (OA/OARG+MKSVM); 3a/b = semidefinite programming approach (OA/OARG+SDP); 4a/b = KCCA approach (OA/OARG+KCCA)

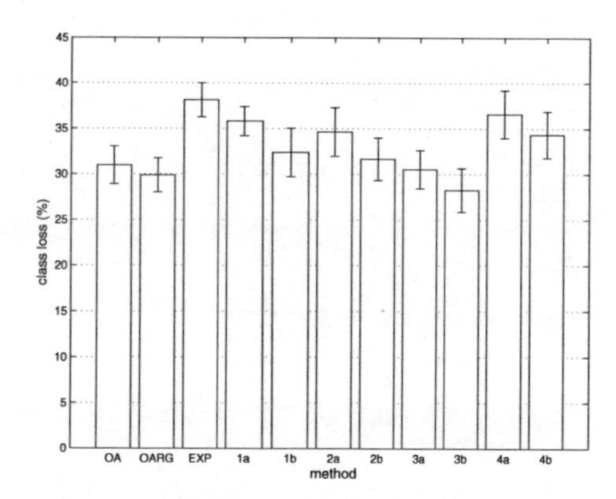

Figure 6.14: Effect of incorporating expert knowledge: 10-fold cross-validation results (class loss in %) for the Yoshida dataset. 1a/b = sum of OA/OARG with EXP model (OA/OARG+EXP); 2a/b = multi-kernel SVM (OA/OARG+MKSVM); 3a/b = semidefinite programming approach (OA/OARG+SDP); 4a/b = KCCA approach (OA/OARG+KCCA)

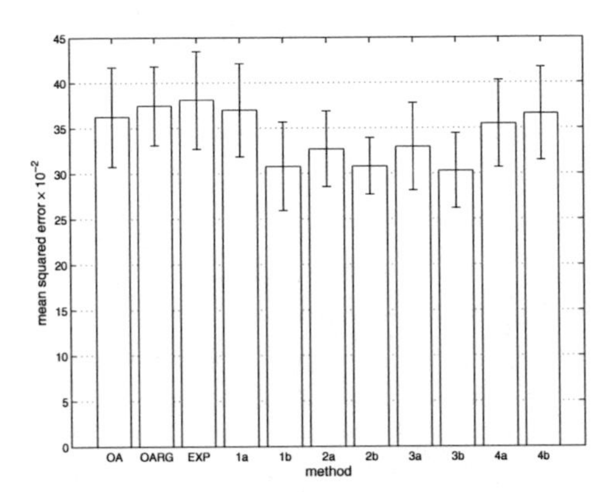

Figure 6.15: Effect of incorporating expert knowledge: 10-fold cross-validation results (mean squared error) for the BBB dataset. 1a/b = sum of OA/OARG with EXP model (OA/OARG+EXP); 2a/b = multi-kernel SVM (OA/OARG+MKSVM); 3a/b = semidefinite programming approach (OA/OARG+SDP); 4a/b = KCCA approach (OA/OARG+KCCA)

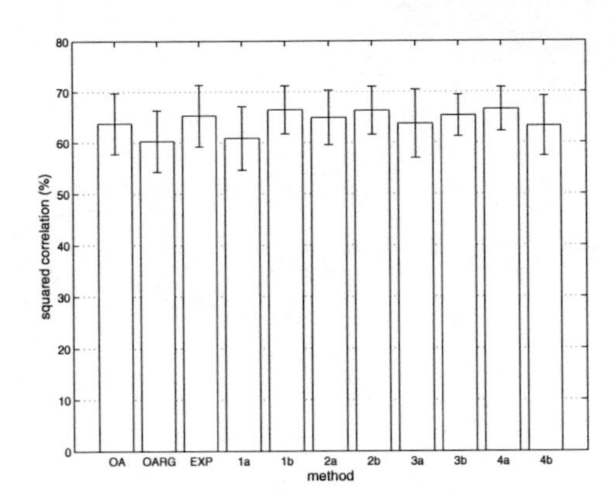

Figure 6.16: Effect of incorporating expert knowledge: 10-fold cross-validation results (squared correlation in %) for the BBB dataset. 1a/b = sum of OA/OARG with EXP model (OA/OARG+EXP); 2a/b = multi-kernel SVM (OA/OARG+MKSVM); 3a/b = semidefinite programming approach (OA/OARG+SDP); 4a/b = KCCA approach (OA/OARG+KCCA)

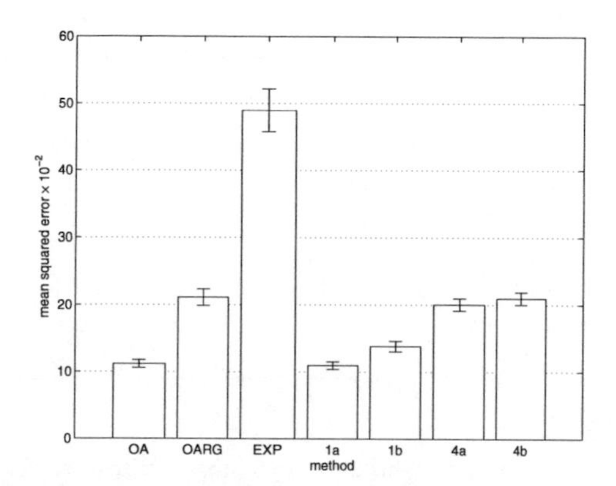

Figure 6.17: Effect of incorporating expert knowledge: 10-fold cross-validation results (mean squared error) for the Huuskonen dataset. 1a/b = sum of OA/OARG with EXP model (OA/OARG+EXP); 2a/b = multi-kernel SVM (OA/OARG+MKSVM); 3a/b = semidefinite programming approach (OA/OARG+SDP); 4a/b = KCCA approach (OA/OARG+KCCA)

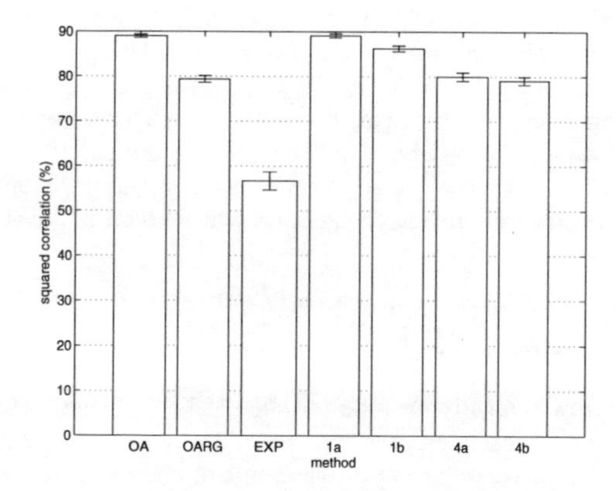

Figure 6.18: Effect of incorporating expert knowledge: 10-fold cross-validation results (squared correlation in %) for the Huuskonen dataset. 1a/b = sum of OA/OARG with EXP model (OA/OARG+EXP); 2a/b = multi-kernel SVM (OA/OARG+MKSVM); 3a/b = semidefinite programming approach (OA/OARG+SDP); 4a/b = KCCA approach (OA/OARG+KCCA)

times for the optimal assignment kernel on average were below those for the marginalized graph kernel. They ranged from around 1.5 to 5ms on average per kernel function evaluation.

Our ad-hoc definition of SMARTS patterns for the reduced graph representation lead to comparable results with the original optimal assignment kernel, except for the Huuskonen dataset. Here our SMARTS patterns seemed not to reflect the chemical background in the right way. The computation times for the OARG kernel were only slightly higher than those for the usual OA kernel.

We investigated the following four strategies to incorporate expert knowledge on relevant descriptors into our approach:

1. an unweighted sum of the optimal assignment kernel and the RBF kernel for the descriptors.
2. a weighted sum, where weights are learned together with the dual variables of the SVM via semidefinite programming.
3. a weighted sum for each molecule separately, which leads to multi-kernel SVMs.
4. correlated features from the feature spaces induced by both kernels, extracted by Kernel CCA.

Our results indicate that especially the first three techniques in most cases lead to a reduction of the error rate, which in some cases was significant. Thereby there does not seem to be a clear winner. For Kernel CCA our results were not as positive. While it never yielded a clear improvement, on the Huuskonen dataset with the OA kernel a significant loss of prediction performance was observed. It seems that in our case the common structure of the feature spaces induced by the graph representation and by the representation with relevant descriptors did not provide enough problem relevant information to reduce the error rate.

6.8 Conclusion

We introduced a new similarity measure for chemical compounds based on a representation of molecules as labeled graphs. The basic idea of our *optimal assignment kernel* is to compute an optimal assignment of the atoms of one molecule to those of another one, including information on neighborhood, membership to certain structural elements and other characteristics. The optimal assignment kernel can be computed efficiently by means of existing algorithms, such as the *Hungarian method* (Kuhn, 1955) or the algorithm described in Mehlhorn and Näher (1999). We showed how the inclusion of neighborhood information for each single atom can be done efficiently via a recursive update equation, even if not only direct neighbors are considered. Comparisons to the marginalized graph kernel by Kashima et al. (2003) in several cases showed a significant reduction of the prediction error on our QSAR/QSPR problems. At the same time our results were at least as good as a classical descriptor model with state-of-the-art SVM descriptor selection. Thereby it is important to point out that in contrast to such a model there was no data dependent adaption of the kernel. We think that this is a special benefit of the kernel approach as it guarantees a unified, highly flexible, easy and fast way to obtain reliable QSAR/QSPR models. However, for the future it might also make sense to perform some form of feature selection among a given set of atomic properties. This can in principle be solved by using e.g. the RFE algorithm. Compared to marginalized graph kernels all in all we see the main advantage of our method that it better reflects a chemist's intuition on the similarity of molecules.

We investigated two major extensions of our approach: the usage of a reduced graph representation, in which certain structural elements are collapsed into a single node of the molecular graph and hence allow to view molecules at different user-specified levels of resolution, and the incorporation of descriptor information known to be relevant for the QSAR/QSPR problem at hand. We showed that the latter can lead to a further significant reduction of the error rate in comparison to the usual optimal assignment kernel, whereas the major benefit of the reduced graph representation lies in the fact that expert knowledge on important structural features can be included.

For the future, the optimal assignment kernel could also give the opportunity to deduce *pharmacophores*. Especially for this purpose the reduced graph representation is advantageous as well. Another open issue related to the reduced graph representation is how an optimal reduced graph could be found by looking at the data. This task certainly involves the necessity to look at the physico-chemical properties of the individual atoms. However, SMARTS patterns currently do not allow the specification of queries including physico-chemical properties. Therefore, more sophisticated ways of defining reduced graphs are required.

Appendix: Preventing the "Tottering"

If we evaluate R_0 (Eq.: 6.11) at all neighbors of a certain topological distance ℓ, we also revisit atoms and bonds that we have considered at topological distance $\ell - 1$. To prevent this "tottering", we can make our decay factor γ dependent not just on the topological distance, but also on the path of visited atoms and bonds. Thereby we have to explicitly forbid paths of the form $a \rightarrow n_i(a) \rightarrow a$. This can be achieved by setting

$$\gamma'(\ell, a, a', n_i(a), n_j(a')) = \begin{cases} 0 & \exists k : n_k(n_i(a)) = a \vee \exists t : n_t(n_j(a')) = a' \\ \gamma(\ell) & \text{otherwise} \end{cases} \quad (6.26)$$

Table 6.3: Effect of the OA kernel with prevention of the "tottering" (OAopt) compared to original OA kernel. For the regression datasets the squared correlation (%) is reported in the second row.

Method	HIA	Yoshida	BBB	SOL	Huuskonen
OA	15.26 ± 2.74	30.98 ± 2.07	36.24 ± 5.48 63.75 ± 6.02	1.01 ± 0.52	11.17 ± 0.6 88.98 ± 0.37
OAopt	15.92 ± 2.64	32.46 ± 2.63	37.79 ± 6.11 62.95 ± 6.39	1.34 ± 0.74	11.71 ± 0.53 88.45 ± 0.31

This requires the following changes in our computation for k_{nei}:

$$k_{nei}(a,a') = k_{atom}(a,a') + \frac{1}{|a||a'|}\sum_{i,j} r_0(a,a',n_i(a),n_j(a')) \tag{6.27}$$

$$+ \sum_{\ell=1}^{L}\left(\frac{1}{|a||a'|}\sum_{i,j}\gamma'(\ell,a,a',n_i(a),n_j(a'))\cdot r_\ell(a,a',n_i(a),n_j(a'))\right)$$

$$r_\ell(a,a',n_i(a),n_j(a')) = \frac{1}{|n_i(a)||n_j(a')|}\cdot \tag{6.28}$$

$$\sum_{k,t} r_{\ell-1}(n_i(a),n_j(a'),n_k(n_i(a)),n_l(n_j(a')))$$

$$r_0(a,a',n_i(a),n_j(a')) = k_{atom}(n_i(a),n_j(a'))k_{bond}(a \to n_i(a), a' \to n_j(a'))$$

That means we can compute k_{nei} by iteratively revisiting the direct neighbors and indirect neighbors of (a,a'). In contrast to (6.12) in (6.27) we do not use an optimal assignment kernel for the direct neighbors of (a,a'), but compute the average match here. Experimentally the modified kernel never leads to an improvement of the prediction error (Table 6.3).

Chapter 7

SVM-Based Detection of Relevant Features that Trigger Action Potentials in Cortical Neurons *in Vivo*

7.1 Introduction

The human brain is the most complex information processing device known to date. Its basic computational units are neural cells which dynamically transform synaptic inputs into action potentials (APs) by intrinsically nonlinear processes. These processes enable networks of neurons to perform complex computations (McCulloch and Pitts, 1943; Minsky and Papert, 1969; Hopfield, 1982). Classically, neurons are viewed as integrators, summing synaptic inputs and emitting an AP once the neuron's membrane potential (MP) reaches a threshold voltage. This reductionist view was extended by Hodgkin and Huxley (1952) who developed a biophysical theory, which could explain the generation of APs. They showed that the emission of an AP is a non-linear high-dimensional process involving the detailed dynamics of voltage-dependent channels in the membrane of a neuron. Although major advances were achieved in describing the dynamics of the underlying voltage-gated channels individually in great detail, emergent mechanisms, which result from their dynamical interplay, were investigated only recently (Koch et al., 1995; y Arcas et al., 2003). In these studies it was predicted that the voltage at which an AP initiates partially depends on the velocity with which the MP depolarizes. This prediction was confirmed in recent experiments, in which the generation of APs in cortical neurons was investigated. They demonstrated that APs were initiated at a low voltage when the MP depolarized rapidly and at a high voltage when the MP depolarized slowly (Azouz and Gray, 2000).

From a functional point of view, the mechanism of AP initiation has a strong qualitative impact on the information processing in the brain. In simplified models it was demonstrated that seemingly minor details of the AP generating mechanism can fundamentally alter the nature of the encoding of synaptic inputs into sequences of APs, as shown in Fourcaud-Trocme et al. (2000) and Naundorf et al. (2004).

Neurons *in vivo* are subject to an immense synaptic bombardment, leading to large fluctuations of their MP (see Fig. 7.1). Although the functional role of these fluctuations is still unclear, it significantly alters the neuron's dynamical properties (Destexhe and Pare, 1999; Volgushev and Eysel, 2000). In the work presented in this Chapter we analyzed the AP generation in this "natural environment" and examined, which patterns in the sub-threshold fluctuations of the MP could predict best the occurrence of an AP. For this purpose we investigated 11 *in vivo* recordings of neurons from cat visual cortex and trained a SVM to discriminate between time points in a recording, which are prior to an AP within a small window, and others, which are not close to an AP (c.f. Fröhlich et al., 2005a). We defined a set of 15 features, which we computed for every time point in a recording, and then applied a RFE-like algorithm to simultaneously select the features in all recordings that contributed most to the classification, without making any a-priori assumptions about the relevance of certain features. This way, we found that the AP initiation can be explained qualitatively best by an interplay of several factors: an increase of the MP several ms (here: 2.5ms) before AP onset, the value and the MP rate of change ("velocity") of the MP just before AP onset, and the mean MP over a longer time interval (here: 50ms). Thereby the feature with the largest impact was the velocity of the MP just before AP onset. In conclusion our results imply that the AP generation mechanism in cortical neurons *in vivo* acts rather as a coincidence detector than as an integrator.

In the next Section we will describe our method in detail. In Section 4.3 we present the results of the application of our method on *in vivo* recordings from cat visual cortex. In Section 4.5 we summarize our results and conclude.

7.2 Our Method

7.2.1 *In Vivo* Recordings

We analyzed 11 *in vivo* recordings from cat visual cortex, where each recording contained at least 100 APs. The recordings exhibited only spontaneous activity generated as the result of the activity of the surrounding network, which lead to low firing rates of less than 2Hz. The experimental details of the data acquisition can be found in Volgushev et al. (2003).

Each recording consists of a discrete time series $T_V = \{(t_i, V(t_i))\}_{i=1}^m$ of length m with MPs $V(t_i)$ and time points t_i. Seven datasets were recorded at a time resolution of 0.1ms, and four at 0.05ms.

7.2.2 Feature Construction

The MP at each time point t_i was embedded into a 15-dimensional vector space, resulting in a vector $\mathbf{x}(t_i)$. The single coordinates of the feature space were defined as:

- the MP $V(t_i)$
- the 1st and the 2nd "derivatives" of the MP, approximated by differences $\Delta V(t_i) := V(t_i) - V(t_{i-1})$ and $\Delta^2 V(t_i) := \Delta V(t_i) - \Delta V(t_{i-1})$
- the mean MPs over 2.5, 5, 25, 50ms before t_i
- the mean of the 1st and 2nd "derivatives" over 2.5, 5, 25, 50ms before t_i

To achieve a higher robustness against noise, we averaged $V(t_i), \Delta V(t_i)$ and $\Delta^2 V(t_i)$ over 3 subsequent points.

The mean MPs over 2.5, 5, 25, 50ms allow a certain separation between "medium time" means and "long time" means. Together with the defined $V(t_i), \Delta V(t_i)$ and $\Delta^2 V(t_i)$ in conclusion we captured "short time", "medium" and "long time" effects by our defined features.

For the construction of the features, we only considered time points lying between the onset of an AP and the end of the preceding one. Thereby, an AP onset was detected by looking for the last time point prior to an AP, for which the MP rate of change was below 1mV/ms. The end of an AP was assumed to be at latest 2ms after the occurrence of the AP maximum.

7.2.3 Definition of a Labeled Dataset

Our goal was to learn the separation of time points in a MP recording, which are $\tau_j$ms prior an AP (with τ_j being some parameter depending on the jth AP that we discuss below) and others that are not close to an AP. Points of the first category are called "positive examples" (or just "positives") in the following, and points of the second one "negative examples" (or just "negatives"). As negative examples we considered all time points having a distance of at least 10ms to the following AP onset (and following the restriction for the feature construction described above — see Fig. 7.1). As the set of positive examples we chose all points in a recording ending at time points $\tilde{t}_j - \tau_j$, where $\tilde{t}_j$ are all those time points of AP maxima that have a time distance of at least 5ms to the preceding one (Fig. 7.2). This procedure reliably removed all burst APs (APs following on each other in an extremely short time interval).

Obviously, for small distances τ_j the set of positive and negative examples can be trivially distinguished by the value of the MP, because we are within an AP. Hence, to ensure a non-trivial classification, τ_j was initialized to the time of the jth AP onset and

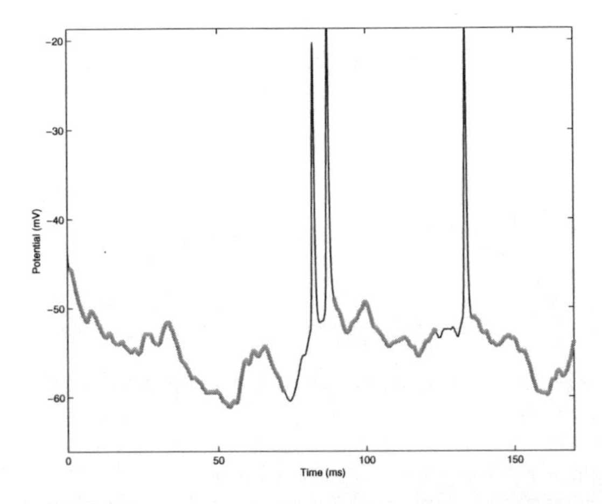

Figure 7.1: Small piece of an example trajectory with negative examples marked thicker.

then further increased in steps of 0.1ms until the distribution of the positives and negatives could not be trivially separated any more with regard to the value of the MP. This was assumed to be true, if the means of both distributions were closer than twice the sum of their standard deviations.

7.2.4 Sub-sampling of Negatives

Compared to the whole length of a trajectory T_V an AP is a rare event. This implies that in a recording we have only few positive examples compared to a huge number of negatives (approx. 10^5 as many as positives — see Fig. 7.3). Highly imbalanced datasets as this always imply a problem for a classifier. First, depending on the overall size of the dataset the computational burden can become quite high. Second and much worse, a SVM trained with the usual hinge loss function can achieve a very high classification accuracy, if just **all** examples are assigned to the negative class. That means we have a true negative rate of 100%, but a true positive rate of 0%. As in our application the number of negative examples is not a limiting factor, we chose to subsample from them. Thereby we concentrated on those negatives, which contained the most information on the classification, i.e. which were most similar to positive ones and thus likely could become support vectors later on. To speed things up, the sub-sampling of the negatives was performed in three stages: First, all negatives were collected that lay within a range

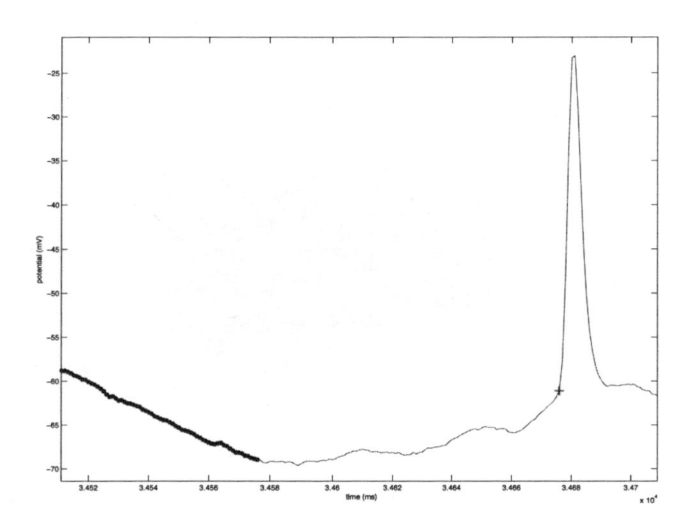

Figure 7.2: A single AP and the positive training example marked as cross just before the AP onset. Further away from the AP the negative training examples are shown as thick dots.

of 2 standard deviations of the distribution of the positives. This excluded all negatives that lay outside the distribution of the positives and hence were trivially distinguishable. Second, among all collected negatives a random subsample of at most 100 times the number of positives was taken. Third, a ν-SVM was trained with the sampled negatives and the positive examples. The ratio behind this method is that the ν-parameter provides a lower bound on the fraction of support vectors (c.f. Chapter 3.2). Hence, by setting $\nu = 2p/(p+n)$ (p = number of positives, n = number of negatives) we get at least $2p$ support vectors of which around half should be negative examples. This firstly leads to a much better balanced dataset and secondly it automatically produces those examples, which we should concentrate on during training (Fig. 7.4, 7.5). Note that we are not interested in the decision function itself here, but just in the selected support vectors.

7.2.5 Selection of Relevant Features

To find out the features that contributed most to the classification out of all 15 ones, we employed a RFE-like algorithm: For each recording we trained a C-SVM with RBF-

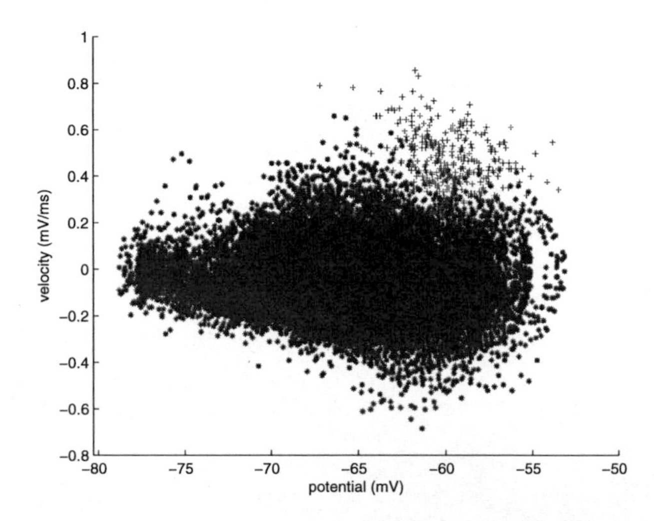

Figure 7.3: Scatter plot of positive (+) and negative examples (dots) in the potential - velocity space. The number of negatives exceeds the number of positives by several magnitudes.

kernel and performed model selection for parameters C and σ (σ = width of RBF kernel) using 5-fold cross-validation. Thereby C was chosen from the interval $[2^{-2}, 2^{14}]$ and σ from $[\hat{\sigma}/4, 4\hat{\sigma}]$, where $\hat{\sigma}$ was set such that $\exp(-D/2\hat{\sigma}) = 0.1$ (D = dimensionality of the data). We then successively eliminated that feature, for which the relative influence on the margin, summed over all recordings, was minimal. The relative influences of the features were determined by the gradient of $\|\mathbf{w}\|^2$ with respect to scaling factors θ of the features (Eq.: 5.4) and then divided by the 1-norm of that gradient.

At each step of the algorithm we recorded the classification loss on the current training set. At the end we returned the feature set, which induced the minimal 5-fold cross-validated classification loss averaged over all recordings.

7.2.6 Evaluation

We evaluated our obtained model using 10-fold cross-validation on each single recording separately. Thereby each actual training set was used only to perform sub-sampling of the negative examples and to perform model selection for the SVM as described above. We computed the relative influence of each feature on the SVM decision function, and

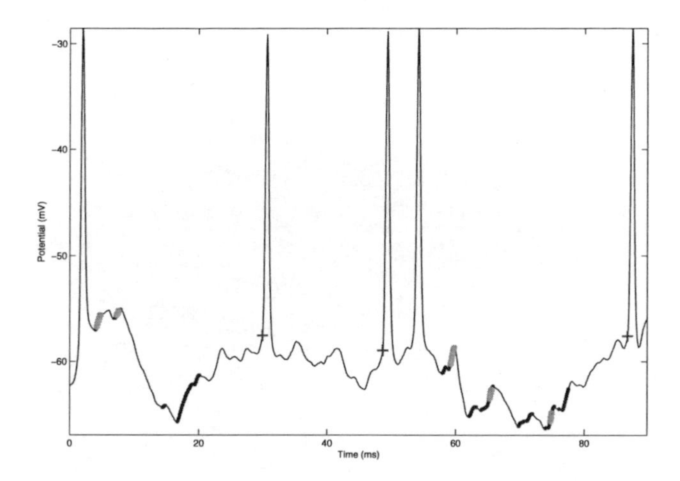

Figure 7.4: Positive and negative example in a MP trajectory. Positive examples are marked with crosses, negative examples are marked thicker. Negative support vectors, which are selected by the ν-SVM sub-sampling method are marked with big dots, which have a lighter color. Note that the third AP is neglected, because it follows the previous AP too closely.

the false negative and the false positive rates averaged over all cross-validation folds. Furthermore, we calculated the *balanced error rates* (BER) as the mean over false positive and false negative rates.

7.3 Results

We applied our method to 11 *in vivo* recordings of spontaneous activity from cat visual cortex. Table 7.1 shows an overview of the different datasets with respect to the number of APs, the number of negative examples in the whole recording, the number of sub-sampled negatives, and the range of the selected distances τ_j from AP maxima. As one can see, the number of sub-sampled negatives is now on a similar scale as the number of positive examples. Furthermore, it is shown that the τ_j for most recordings are between 0.4 and 0.7ms, whereas for the first and the last one they are above 1.1ms. The reason for this is that the magnitude of the sub-threshold fluctuations for the individual recordings differ.

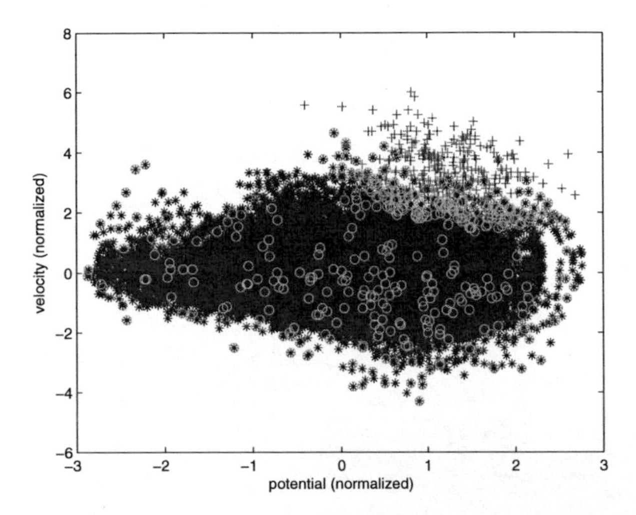

Figure 7.5: Scatter plot of positive (+) and negative (stars) examples in a normalized potential-velocity space: Sub-sampled negative support vectors are marked by extra circles.

As a proof of concept in Table 7.2 we demonstrate the inability of a SVM to learn the discrimination of positive and negative examples just based on the value of the MP for the given τ_j. The results indicate a performance worse than chance level. In contrast, Table 7.3 for the model using all 15 features shows a very low BER in all cases. At the same time, the false positive rate is much less than 1%, which demonstrates the success of the sub-sampling strategy.

Applying the feature selection method described in 7.2.5 we obtained the following 7 features:

- the MP $V(t_i)$
- the 1st "derivative" of the MP
- the mean MPs over 2.5 and 50ms
- the mean of the 1st "derivatives" over 2.5, 25 and 50ms

Compared to the usage of all features the results (Table 7.4) in 4 cases lead to significant worse false positive rates (*fpr*), in one case to a significant worse false negative rate (*fnr*) and in one case to a significant better false negative rate. Thereby significance was tested by a two-tailed paired t-test at 5% significance level. At the same time the BER was only significantly lower in one and worse in one other situation. Thus, all in all the results seem to be roughly comparable to the usage of all features.

However, a more detailed analysis of the selected features reveals that the "long time" means of the 1st "derivative" should to be correlated with the MP $V(t_i)$, because

$$\frac{1}{N}\sum_{j=0}^{N}\Delta V(t_{i-j}) = \frac{1}{N}(V(t_i) - V(t_{i-N})) \tag{7.1}$$

For N being large enough the time point t_{i-N} should be quite far away from t_i, especially for t_i being the time point of a positive example. Hence, we can describe $V(t_{i-N})$ by a normally distributed random variable with mean equal to the average MP within all negative trajectories. Therefore, the "long time" mean of the 1st "derivative" is effectively a linear function of the MP $V(t_i)$ with an additive noise term.

If we assume $N = 25$ms to be sufficiently large, this suggests the elimination of the "long time" means of the 1st "derivative" from our feature set. As shown in Table 7.5 compared to the usage of all features we now achieve a significant improvement of the false positive rate and the false negative rate in 2 cases and a loss in just one case each. The BER is significantly better in one case and worse in another one, as before. All in all the results seem to be comparable to those with 7 features and to the usage of all features.

The relative influences of the 5 selected features on the decision function, averaged over all recordings, are depicted in Fig. 7.6. They were computed as described in Subsection 7.2.5. As one can see the influence of the 1st "derivative" is highest, whereas that of the mean potential over 2.5ms and the "long time" mean of the MP is significantly lower.

Interpreting the selected features, we can identify four basic factors that according to our experiments should determine the generation of an AP in cortical neurons *in vivo*:

1. the value of the MP just before AP onset (feature: $V(t_i)$)
2. the MP rate of change just before AP onset (features $\Delta V(t_i)$)
3. an increase of the MP over a short period before AP onset (features: mean MP and 1st "deriv." over 2.5ms)
4. the long time value of the MP (feature: mean MP over 50ms)

7.4 Conclusion

We investigated the AP generation in *in vivo* recordings from cat visual cortex. We introduced a machine learning based method to work out, which features of the MP are

Table 7.1: Overview of the different in-vivo recordings used in our experiments.

Recording	#AP	#Negatives	#Negative SVs	τ_j(ms)
1	121	213300	455	1.1 - 1.3
2	217	39208	588	0.6 - 0.8
3	340	27596	995	0.4 - 0.7
4	300	32270	895	0.4 - 0.7
5	127	53374	659	0.4 - 0.5
6	202	44245	729	0.4 - 0.5
7	136	1361480	611	0.5 - 0.6
8	1155	2930116	1895	0.5 - 0.7
9	205	2447870	822	0.5 - 0.7
10	393	3848122	927	0.5 - 0.7
11	193	1374508	956	1.1 - 1.4
Average	308	1124735	867	0.7

Table 7.2: 10-fold CV classification results $\pm$ std. err. using the **value of the potential** only.

Recording	fnr (%)	fpr (%)	BER (%)
1	32.18 ± 13.66	97.95 ± 0.37	65.06 ± 6.97
2	61.15 ± 13.61	80.24 ± 2.69	70.69 ± 5.88
3	32.94 ± 11.91	87.08 ± 1.61	60.01 ± 5.66
4	48.00 ± 13.26	87.21 ± 3.67	67.64 ± 5.02
5	59.29 ± 18.58	88.75 ± 1.58	74.02 ± 8.68
6	47.88 ± 16.06	90.27 ± 1.46	69.08 ± 7.82
7	70.33 ± 17.56	86.11 ± 30.46	78.22 ± 12.49
8	24.07 ± 5.87	99.11 ± 0.15	61.59 ± 2.93
9	64.83 ± 16.21	97.50 ± 0.63	81.17 ± 7.92
10	77.58 ± 18.58	98.43 ± 1.74	88.01 ± 9.6
11	25.89 ± 6.41	98.29 ± 0.42	62.09 ± 3.19
Average	49.56 ± 13.79	91.90 ± 4.08	70.73 ± 6.92

Table 7.3: 10-fold CV classification results ± std. err. using **all** features.

Recording	**fnr (%)**	**fpr (%)**	**BER (%)**
1	7.44 ± 7.29	0.06 ± 0.03	3.75 ± 3.64
2	8.77 ± 4.06	0.03 ± 0.02	4.40 ± 2.03
3	15.59 ± 7.47	0.08 ± 0.06	7.83 ± 3.72
4	16.33 ± 5.97	0.22 ± 0.33	8.28 ± 2.99
5	15.06 ± 7.07	0.08 ± 0.04	7.57 ± 3.54
6	13.86 ± 4.51	0.05 ± 0.03	6.95 ± 2.25
7	6.65 ± 5.48	0.06 ± 0.01	3.45 ± 2.74
8	5.88 ± 2.39	0.06 ± 0.01	2.97 ± 1.19
9	26.29 ± 13.02	0.19 ± 0.02	13.24 ± 6.5
10	3.31 ± 2.10	0.04 ± 0.01	1.67 ± 1.05
11	23.89 ± 9.05	0.1 ± 0.01	12.00 ± 4.53
Average	13.00 ± 6.22	0.09 ± 0.05	6.54 ± 3.10

Table 7.4: 10-fold CV classification results ± std. err. using the consistent model with **7 features**. Significant wins or losses compared to the model with all features are denoted by "*" and "-", respectively.

Recording	**fnr (%)**	**fpr (%)**	**BER (%)**
1	8.27 ± 6.81	0.05 ± 0.01	4.16 ± 3.4
2	7.86 ± 6.22	0.06 ± 0.04	3.96 ± 3.11
3	11.76 ± 4.38	$1.96 \pm 1.88^{-}$	6.86 ± 2.65
4	14.33 ± 5.22	1.12 ± 2.52	7.73 ± 2.97
5	8.78 ± 8.96*	0.06 ± 0.02	4.42 ± 4.48*
6	11.83 ± 3.19	$1.48 \pm 1.21^{-}$	6.66 ± 1.91
7	5.11 ± 6.08	0.06 ± 0.01	2.58 ± 3.04
8	5.88 ± 1.50	$0.99 \pm 1.08^{-}$	3.43 ± 0.80
9	19.88 ± 11.31	0.21 ± 0.06	10.05 ± 5.63
10	$5.08 \pm 2.90^{-}$	0.10 ± 0.10	$2.59 \pm 1.42^{-}$
11	24.39 ± 10.17	$0.12 \pm 0.01^{-}$	12.26 ± 5.08
Average	11.12 ± 6.07	0.55 ± 0.63	5.86 ± 3.14

Table 7.5: 10-fold CV classification results ± std. err. using the consistent model with **5 features**.

Recording	fnr (%)	fpr (%)	BER (%)
1	8.27 ± 6.81	0.06 ± 0.02	4.16 ± 3.40
2	11.56 ± 9.13	0.05 ± 0.03	5.81 ± 4.56
3	$10.88 \pm 8.55^{*}$	$0.7 \pm 0.64^{*}$	5.79 ± 4.34
4	12.67 ± 7.98	0.68 ± 1.90	6.68 ± 4.13
5	13.65 ± 12.08	0.14 ± 0.13	6.90 ± 6.07
6	9.38 ± 7.19	$3.71 \pm 1.91^{-}$	6.55 ± 3.90
7	8.74 ± 5.75	$0.04 \pm 0.01^{*}$	4.39 ± 2.88
8	5.37 ± 1.72	1.29 ± 2.39	3.33 ± 1.69
9	$19.00 \pm 5.72^{*}$	0.29 ± 0.15	$9.64 \pm 2.89^{*}$
10	$5.33 \pm 3.02^{-}$	0.17 ± 0.28	$2.75 \pm 1.53^{-}$
11	24.39 ± 7.37	0.11 ± 0.01	12.25 ± 3.68
Average	11.75 ± 6.85	0.66 ± 0.68	6.21 ± 3.57

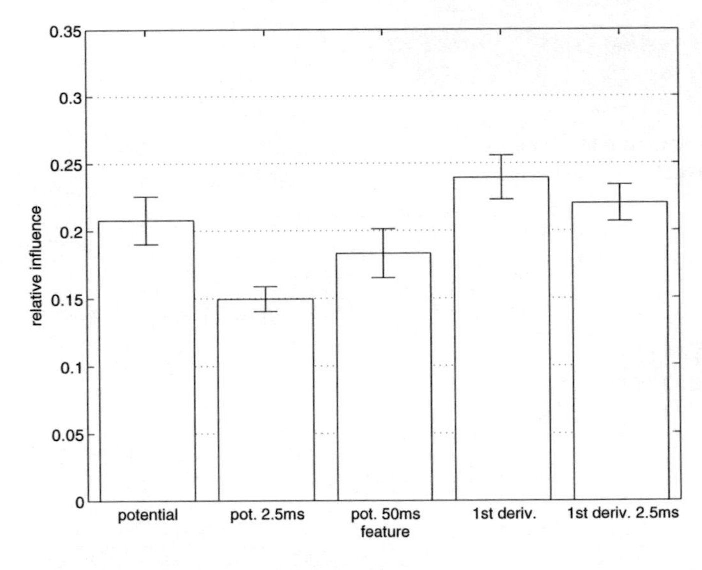

Figure 7.6: Relative influence of the most relevant features on the SVM-decision function averaged over all datasets.

most relevant for the occurrence of an AP. We inferred a SVM model that could classify with high accuracy between time points prior to an AP and others which are not close to an AP. Without making any a-priori assumptions on the relevance of certain features, we used a RFE-like algorithm to detect the features, which are of highest relevance for the classification. This way we found a model that is based on four basic factors: the value and the velocity of the MP just before AP onset, an increase of the MP over a short period before AP onset (here: 2.5ms), and the average value of the MP over a longer time interval (here: 50ms). Using the features describing these factors, the occurrence of an AP could be predicted with very high accuracy on all our recordings. Thereby the feature with the largest impact was the MP rate of change just before AP onset. Our results imply that cortical neurons act as coincidence detectors, which are most sensitive to fast changes of the MP. Models of cortical neurons, which assume a voltage threshold for AP initiation thus might not reflect the dynamics of AP generation in the right way.

Several previous theoretical studies assessed the computation performed in single neurons (e.g. Koch et al., 1995; y Arcas et al., 2003). These studies were almost exclusively based on the Hodgkin-Huxley model (Hodgkin and Huxley, 1952), which describes the dynamical interplay of voltage-gated sodium and potassium channels in the generation of an AP. It is, however, not a-priori clear that this dynamics is equivalent with the AP generation in cortical neurons. Experimentally, the AP generation in cortical neurons *in vivo* was recently addressed in Azouz and Gray (2000). In this study, a correlation between the potential and the first derivative at AP onset was proposed to explain the large variability of AP onset potentials found *in vivo*. Our approach generalizes these results by using a set of 15 features without any a-priori weighting. It therefore allowed for the systematic construction of a simple phenomenological model, which reproduces the AP generation with very high accuracy and serves at a starting point for the development for simplified phenomenological neuron models, which reproduce the dynamical AP initiation of cortical neurons *in vivo*.

Chapter 8

Kernel-Based Functional Grouping of Genes Using the Gene Ontology

8.1 Introduction

In the past few years, DNA microarrays have become major tools in the field of functional genomics. In contrast to traditional methods, these technologies enable the researchers to collect tremendous amounts of data, whose analysis itself constitutes a challenge. On the other side, these high throughput methods provide a global view on the cellular processes as well as their underlying regulatory mechanisms and are therefore quite popular among biologists.

During the analysis of such data, researchers apply different approaches in order to deal with the huge amounts of data they receive. Some use statistics to find significantly regulated genes that due to their change in expression may be involved in the underlying process. Others apply pattern recognition methods to cluster the genes according to their expression profiles (Eisen et al., 1998). The hypothesis is, that genes with expression pattern similarity to known genes involved in the examined biological process, may play a role in the process, too. In both cases, researchers often end up with long lists of interesting candidate genes that need further examination. At this point a second step is almost always applied: biologists categorize these genes to known biological functions and thus try to combine a pure numerical analysis with biological information. In this Chapter we address the problem of finding gene functional clusters solely based on Gene Ontology (GO) terms (The Gene Ontology Consortium, 2004). The advantage of such a method is that no prior knowledge about relevant pathways is necessary except a mapping from genes to their ontological information. The latter is often available in public databases.

While GO analysis is an increasingly important field, existing techniques suffer from some weaknesses: Many methods consider the GO simply as a list of terms, ignoring any structural relationships (Beißbarth and Speed, 2004; Shah and Fedoroff, 2004; Gat-Viks et al., 2003; Robinson et al., 2003; Zeeberg et al., 2003). Others regard the GO primarily as a tree and convert the GO graph into a tree structure for determining distances between nodes (Lee et al., 2004). Again others use a very simple distance measure that relies on counting path lengths (Joslyn et al., 2004). This is a delicate approach in imbalanced graphs like the GO, whose subgraphs have different degrees of detail. Besides, such a distance does not depend on the distribution of terms in the dataset. Therefore, it may not be feasible for clustering, because it may be impossible to resolve different clusters, if the data concentrates on a subgraph of the GO. The aim of many methods is primarily either to use the GO as preprocessing (Adryan and Schuh, 2004) or as visualization tool (Doniger et al., 2003). Only few approaches utilize its structure for computation. Many methods look for statistically overrepresented GO terms in a list of annotated genes (Beißbarth and Speed, 2004; Shah and Fedoroff, 2004; Gat-Viks et al., 2003; Robinson et al., 2003; Zeeberg et al., 2003; Doniger et al., 2003; Lee et al., 2004).

To our best knowledge so far there exists no clustering based method that produces a biologically plausible functional grouping of genes based on the GO structure despite our earlier publications (Speer et al., 2004, 2005a,b,c; Fröhlich et al., 2006a). Probably the work most closely related to ours is that of Joslyn et al. (2004), which does not employ clustering algorithms, but looks for terms in the GO hierarchy that have at minimum a user specified coverage of the genes in the given dataset.

Given the GO annotations of the genes in our dataset we are able to compute a functional similarity between genes, which is provably a symmetric and positive semidefinite kernel function. This kernel function is then used to cluster genes according to their biological function by the spectral clustering of Ng et al. (2002), dual k-means (e.g. Shawe-Taylor and Cristianini, 2004) or average linkage performed in a feature space (c.f. Chapter 3.6). We investigate two ways of constructing a kernel function: one is the usage of the optimal assignment kernel from Chapter 6.2 in combination with a special kernel function comparing individual GO terms, the other approach works by constructing an empirical kernel map (Schölkopf and Smola, 2002) from feature vectors describing for each gene the GO based similarity to certain prototype genes.

The organization of this Chapter is as follows: a brief introduction to the Gene Ontology is given in Section 8.2. Section 8.3 explains our method in detail. The performance of our approach on real world datasets is evaluated in Section 8.4.2. Section 8.5 discusses the results and outlines areas of future research.

8.2 The Gene Ontology

The Gene Ontology (GO) is one of the most important ontologies within the bioinformatics community. It is developed by the GO Consortium (The Gene Ontology Consortium, 2004) and specifically intended for annotating gene products with a consistent, controlled and structured vocabulary. The GO is limited to the annotation of gene products and inde-

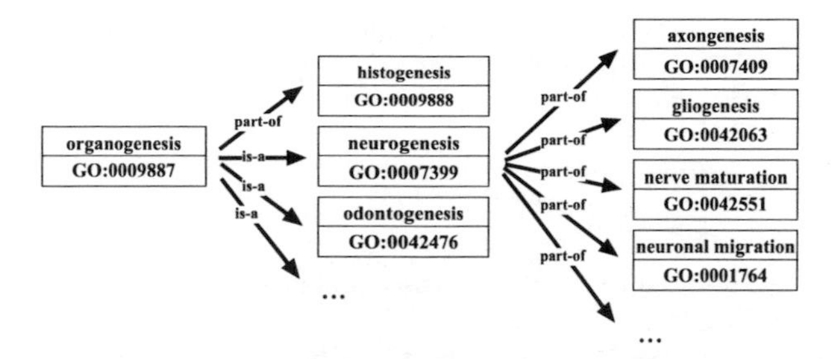

Figure 8.1: Relations in the Gene Ontology. Each node is annotated with a unique accession number.

pendent from any biological species. It is rapidly growing, having about 19,000 terms (as of October 2005) and additionally new ontologies covering other biological or medical aspects are being developed.

The GO represents terms in a Directed Acyclic Graph (DAG) covering three orthogonal taxonomies or "aspects": *molecular function*, *biological process* and *cellular component*. The GO graph consists of a number of terms represented as nodes, which are connected by relationships represented as edges. Terms are allowed to have multiple parents as well as multiple children. Two different kinds of relationship exist: the "is-a" relationship (*neurogenesis* and *odontogenesis* are for example children of *organogenesis*) and the "part-of" relationship which describes e.g. that *histogenesis* is part of *organogenesis* or *axongenesis* is part of *neurogenesis*.

The GO terms are used to annotate gene products in the widest sense, e.g. sequences in databases as well as measured expression profiles. By providing a standard vocabulary across any biological resources, the GO enables researchers to use this information for automatic data analysis. The GO is available as flat files and XML files and has also been ported to a MySQL database scheme (The Gene Ontology Consortium, 2004).

8.3 Kernels for Functional Gene Grouping

8.3.1 Optimal Assignment Kernels for Genes

Our goal is to cluster genes according to their function. Hence, we need some similarity measure, which in our case will be a kernel function. However, the problem when comparing two genes g and g' is that each gene can have multiple functions, i.e. it can be mapped on different nodes in the GO graph. On the other hand each node in the GO graph can correspond to multiple genes. Hence, we need a way to compare g and g' with annotated lists of GO terms $t_1, ..., t_n$ and $t'_1, ..., t'_m$. Assumed we already have a kernel function k_T comparing two GO terms t and t', then a way of comparing g and g' is to assign each term in the smaller of both lists to exactly one term in the longer one such that the sum of term similarities is maximized. This is exactly the optimal assignment kernel from Chapter 6.2:

$$k(g, g') = \begin{cases} \max_\pi \sum_{i=1}^{n} k_T(t_i, t'_{\pi(i)}) & \text{if } n > m \\ \max_\pi \sum_{j=1}^{m} k_T(t_{\pi(j)}, t'_j) & \text{otherwise} \end{cases} \quad (8.1)$$

where π is some permutation of either an n-subset of natural numbers $1, ..., m$ or an m-subset of natural numbers $1, ..., n$.

Like in Chapter 6.2 we should further normalize the kernel to prevent that larger lists of terms automatically achieve a higher similarity (Schölkopf and Smola, 2002)

$$k(g, g') \leftarrow \frac{k(g, g')}{\sqrt{k(g, g)k(g', g')}} \quad (8.2)$$

Finally, the spectrum of the kernel matrix $\mathbf{K} = (k(g_i, g_j))_{i,j=1}^{N}$ calculated on a set of genes $g_1, ..., g_N$ has to be shifted by the smallest negative eigenvalue (if there is any) to make it positive semidefinite (c.f. Chapter 6.2). Let us now turn to the question how the kernel k_T between two terms t, t' can be defined.

8.3.2 Kernel Functions for Gene Ontology Terms

There are a couple of semantic similarity and distance measures of different complexity (Budanitsky and Hirst, 2001), most of them were originally developed for taxonomies like WordNet. Here we use a similarity measure based on the information content of each GO term (Resnik, 1995). The information content of a term is defined as the probability with which this term or any child term occurs in a dataset. Following the notation in information theory, the information content (IC) of a term t can be quantified as follows:

$$IC(c) = -\log P(t) \quad (8.3)$$

where $P(t)$ is the probability of encountering an instance of class t.

In the case of a hierarchical structure, such as the GO, where a term in the hierarchy subsumes those lower in the hierarchy, this implies that $P(t)$ is monotonic as one moves toward the root node. As the node's probability increases, its information content or its informativeness decreases. The root node has a probability of 1, hence its information content is 0. As the three aspects of the GO are disconnected subgraphs, this is still true if we ignore the root node (*Gene Ontology*, GO:0003673) and take, e.g., *cellular component* (GO:0005575) as our root node instead. $P(t)$ is simply computed using maximum likelihood estimation:

$$P(t) = \frac{\text{freq}(t)}{N} \quad (8.4)$$

where N is the total number of terms occurring in the dataset and freq(t) is the number of times term t or any child term of t occurs in the dataset.

The similarity of two terms t, t' can then be defined as follows:

$$k_T(t,t') = -\log \max_{\hat{t} \in Pa(t,t')} P(\hat{t}) = -\log P_{ms}(t,t') \tag{8.5}$$

where $Pa(t,t')$ is the set of parental terms shared by both t and t'. As the GO allows multiple parents for each term, two terms can share parents by multiple paths. We take the minimum $P(\hat{t})$, if there is more than one parent. This is called P_{ms}, for *probability of the minimum subsumer* (Lord et al., 2002). Note that due to the nature of the GO graph topology we have $\forall t, t' : P_{ms}(t,t') \geq P(t), P_{ms}(t,t') \geq P(t')$, because the common ancestor of t and t' must be higher in the graph hierarchy than each of them.

As already mentioned by Ben-Hur and Noble (2005) the function k_T is a valid positive semidefinite kernel, because it defines a dot product with respect to the infinity norm in the annotation space. There is also an interesting connection to a distance measure between two terms t, t' that was developed by Jiang and Conrath (1998):

$$\begin{aligned} d(t,t') &= 2\log P_{ms}(t,t') - (\log P(t) + \log P(t')) \\ &= k_T(t,t) - 2k_T(t,t') + k_T(t',t') \end{aligned} \tag{8.6}$$

8.3.3 Empirical Kernel Maps for Genes

A different type of GO based kernel construction for genes was given by Speer and Fröhlich (Speer et al., 2005a): For each gene g annotated with GO terms $t_1, ..., t_n$ we construct a feature vector $\phi_{\mathbf{p}}(g)$ relative to prototype genes $\mathbf{p} = (p_1, ..., p_N)^T$ where each prototype gene p_i is annotated with GO terms $t'_{i1}, ..., t'_{in_i}$:

$$\phi_{\mathbf{p}}(g) = (\hat{d}_1(g,p_1), ..., \hat{d}_N(g,p_N))^T \tag{8.7}$$

where $\hat{d}_i(g,p_i)$ denotes the **minimal** distance between terms belonging to gene g and prototype p_i. This leads to the so-called *empirical kernel map* (Schölkopf and Smola, 2002)

$$k(g,g') = \langle \phi_{\mathbf{p}}(g), \phi_{\mathbf{p}}(g') \rangle \tag{8.8}$$

That means each gene g is represented by its minimal term distance (8.6) with respect to all prototype genes. In the simplest case prototypes are just all genes from our dataset. The advantage of this approach is that this way more information is incorporated into the kernel. The obvious disadvantage, on the other hand, is that we have a large number of features, which makes this approach impracticable for larger datasets. A possible solution to this problem would be to perform PCA (Hotelling, 1933) on the feature vectors before applying the clustering. It was also suggested to calculate a clustering on the terms itself to reduce their number and this way to achieve a speed-up (Speer et al., 2005b).

Similar to above, one should normalize the feature vectors to norm 1.

8.4 Experiments

8.4.1 Datasets

One possible scenario where researchers like to group a list of genes according to their function is when they examine gene expression with DNA microarray technology, afterwards apply some filtering or statistical analysis and end up with a list of genes that show a significant change in their expression according to a control experiment. Therefore, we chose two publicly available microarray datasets, annotated the genes with GO information and used them for functional clustering.

The authors of the first dataset (Iyer et al., 1999) examined the response of human fibroblasts to serum on cDNA microarrays in order to study growth control and cell cycle progression. They found 517 genes whose expression levels varied significantly, for details see Iyer et al. (1999). We used these 517 genes for which the authors provide NCBI accession numbers. The GO mapping was done via Gene Lynx (2004) IDs. After mapping to the GO, 238 genes showed one or more mappings to *biological process* or a child term of *biological process*. These 238 genes were used for the clustering.

In order to study gene regulation during eukaryotic mitosis, the authors of the second dataset examined the transcriptional profiling of human fibroblasts during cell cycle using microarrays (Cho et al., 2001). Duplicate experiments were carried out at 13 different time points ranging from 0 to 24 hours. Cho et al. (2001) found 388 genes whose expression levels varied significantly. Hvidsten et al. (2003) provides a mapping of the dataset to GO. 233 of the 388 genes showed at least one mapping to the GO *biological process* taxonomy and were thus used for clustering.

8.4.2 Results

We compared clustering results of the empirical kernel map (8.8) with and without performing PCA preprocessing (EKM/EKM+PCA) and the optimal assignment kernel combined with the term kernel (8.5) (OA-Term). Clusterings were computed by the spectral clustering algorithm of Ng et al. (2002), by (dual) k-means (e.g. Shawe-Taylor and Cristianini, 2004) and average linkage. Thereby, for the k-means we used 50 restarts with random initialization and returned the clustering with the minimal distortion. Clusters, which became empty during the optimization process, were reinitialized with the point being furthest from its centroid.

In case of the feature vector representation (8.7) for the spectral clustering we employed the usual embedding via an RBF kernel of width σ. The parameter σ was tuned in the range $\hat{\sigma}/4, ..., 4\hat{\sigma}$ such that the relation between the within-cluster-distortion and the between-cluster-distortion became minimal (c.f. Duda et al., 2001; Ng et al., 2002). Thereby we set $\hat{\sigma}$ such that $\exp(-D/(2\hat{\sigma}^2)) = 0.01$ (D = dimensionality of the feature vector).

For the PCA preprocessing we extracted as many principal components such that at least 95% of the total variance could be explained. This lead to just 17 principal components for the first and 15 principal components for the second dataset.

To validate our results we plotted the silhouettes for 15 clusters and computed the mean silhouette index (Rousseeuw, 1987). The silhouette value for each point is a measure of how similar that point is to points in its own cluster vs. points in other clusters, and ranges from -1 to +1. It is defined as:

$$S(i) = \frac{\min_j(\bar{d}_B(i,j)) - \bar{d}_W(i)}{\max(\bar{d}_W(i), \min_j(\bar{d}_B(i,j))} \tag{8.9}$$

where $\bar{d}_W(i)$ is the average distance from the i-th point to the other points in its own cluster, and $\bar{d}_B(i,j)$ is the average distance from the i-th point to points in another cluster j.

As one can see from Table 8.1 the mean silhouette indices for the empirical kernel map are consistently better than those for the OA-Term method. Thereby the EKM+PCA approach always gives the best results. The silhouette plots depicted in Fig. 8.2 and 8.3 show that especially average linkage, but also (less extreme) spectral clustering tend to form very large clusters with high diversity, whereas the clusters obtained by k-means are more balanced.

In Tables F.1 to F.5 (Appendix) we show some example clusters found by dual k-means in combination with our kernels. Table F.1 shows a cluster that was identically found by both EKM based methods (with and without PCA), whereas Table F.2 shows the corresponding cluster that was found using the optimal assignment kernel. Both clusters contain genes that are annotated with GO terms related to metabolism, but the cluster found by the optimal assignment kernel additionally contains genes related to RNA processing (marked bold). Another example are the clusters shown in Tables F.3 to F.5 found by the OA-Term method (Table F.3) and the EKM with and without PCA (Tables F.4 and F.5). Again, both EKM approaches produce clusters that only contain genes related to DNA replication, whereas the cluster produced by the optimal assignment kernel contains also genes related to transcription (marked in italics) and energy pathways (marked in bold).

8.5 Conclusion

We introduced symmetric and positive semidefinite kernel functions to assess the functional similarity of genes using the Gene Ontology only. These kernel functions can be used in combination with any kernel-based clustering algorithm such as dual k-means to automatically group genes with regard to their function. We found that the kernel function based on an empirical kernel map combined with a PCA preprocessing of the feature vectors leads to better clustering results than without PCA preprocessing and also better

Table 8.1: Mean silhouette indices for the empirical kernel map without (EKM) and with PCA preprocessing (EKM+PCA) and the optimal assignment kernel combined with term kernel (8.5) (OA-Term). "spectral" = spectral clustering; "AL" = average linkage

Dataset	Method	EKM	EKM+PCA	OA-Term
	spectral	0.40 ± 0.27	$\mathbf{0.50 \pm 0.25}$	0.23 ± 0.19
Iyer et *al.*	k-means	0.40 ± 0.23	$\mathbf{0.50 \pm 0.24}$	0.16 ± 0.15
	AL	0.38 ± 0.24	0.47 ± 0.26	0.24 ± 0.21
	spectral	0.46 ± 0.24	$\mathbf{0.54 \pm 0.24}$	0.22 ± 0.21
Cho et *al.*	k-means	0.40 ± 0.24	0.50 ± 0.25	0.16 ± 0.17
	AL	0.33 ± 0.24	0.49 ± 0.25	0.21 ± 0.21

results than the optimal assignment kernel in combination with a kernel comparing pairs of GO terms. When evaluating the GO annotations in detail, both empirical kernel map based kernels seem to lead to more compact clusters than the optimal assignment kernel. This can be explained by the richer representation utilized in the empirical kernel map. Nevertheless, the more direct optimal assignment kernel approach has the advantage that much fewer features need to be calculated explicitly and hence this method might be better suited for large datasets. It remains an open issue to test this systematically in practice.

All in all, we showed that all three methods are able to detect functional clusters of genes. We thus think that finally it depends on the problem at hand and the size of the dataset, which one is favorable.

In principle, a syntactical clustering approach as presented here could be employed for other ontologies as the GO as well. In fact, most ontologies are organized in structures that are similar to the GO (think e.g. about an ontology of species in zoology). Thus our method could provide a useful tool for handling data coming from such a domain in an automated way and help humans to get a better understanding of it.

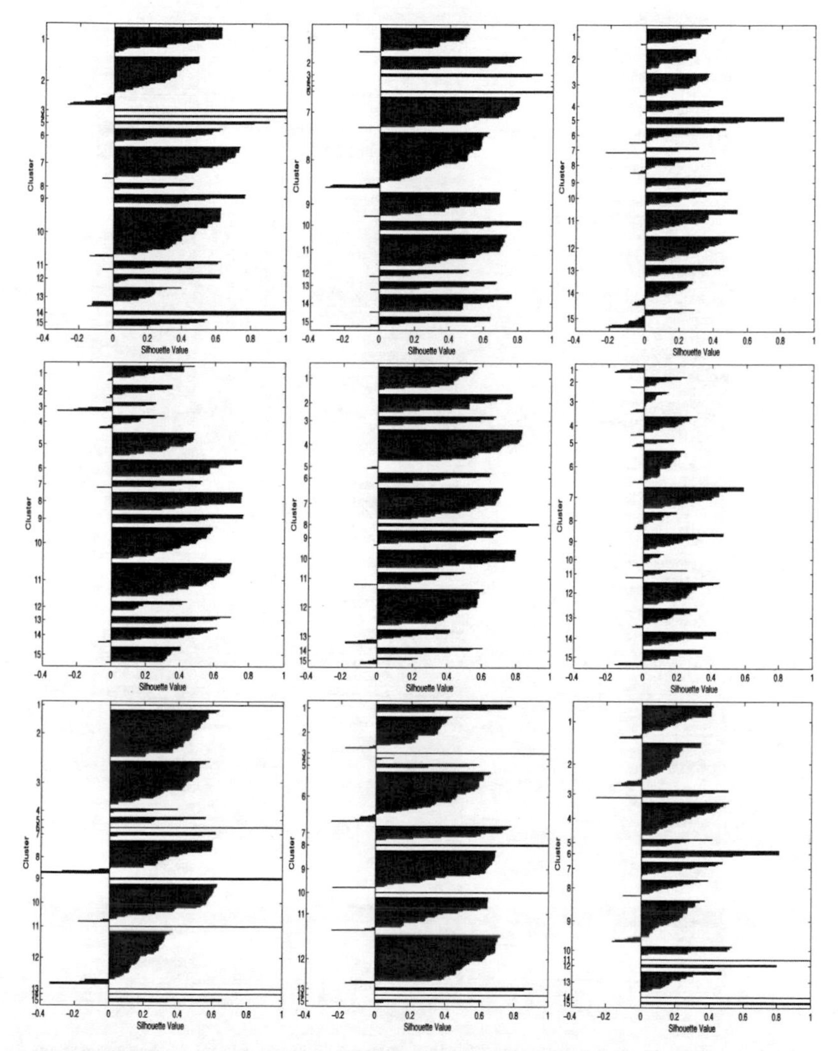

Figure 8.2: Silhouette plots obtained by the spectral clustering (row 1), k-means (row 2) and average linkage (row 3) using EKM (first column), EKM+PCA (second column) and OA-Term (third column) for the first dataset.

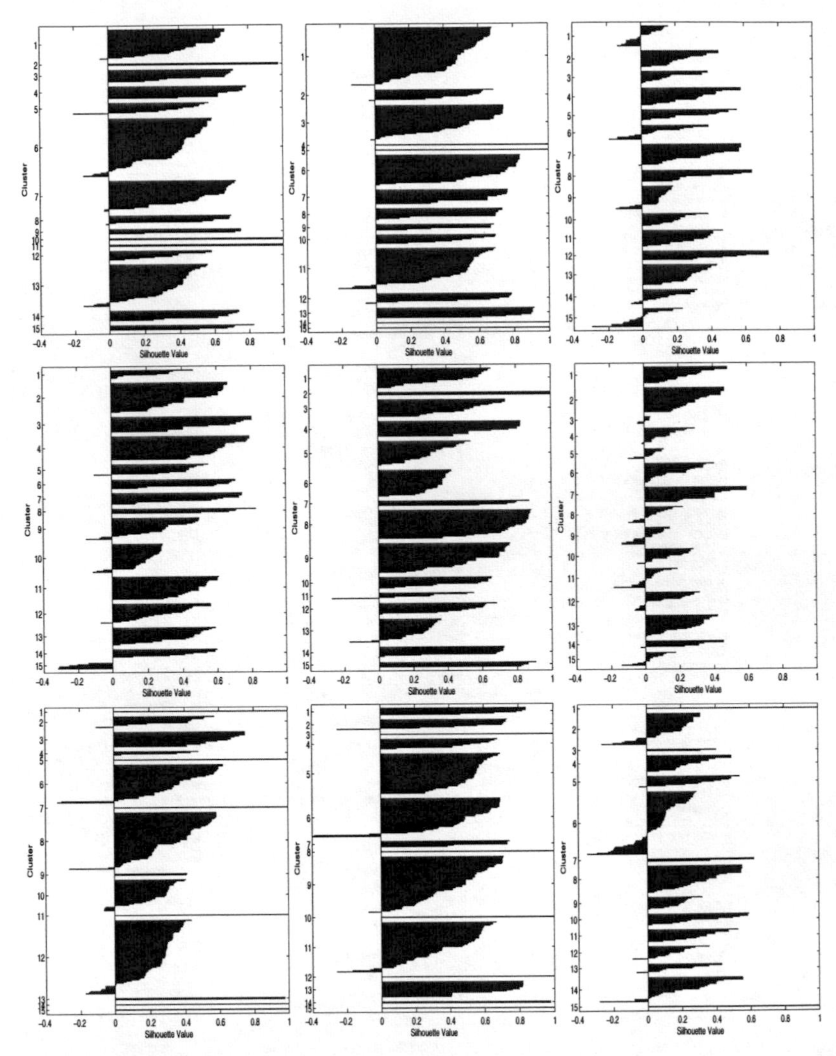

Figure 8.3: Silhouette plots obtained by the spectral clustering (row 1), k-means (row 2) and average linkage (row 3) using EKM (first column), EKM+PCA (second column) and OA-Term (third column) for the second dataset.

Chapter 9

Summary

We briefly summarize the main results of this thesis: In Chapter 4 we presented a new model selection algorithm for SVMs in classification and regression, which is based on the idea to learn an Online Gaussian Process of the error surface of the SVM in parameter space. A new parameter vector is sampled according to the expected improvement criterion (Jones et al., 1998). The algorithm can deal with arbitrary kernel functions, even if they are not differentiable. On a typical classification experiment with a RBF kernel only around 10% of the number of search steps of a grid search are needed, on a typical regression experiment only around 1%. At the same time the prediction performance is comparable. Experimental comparisons of our algorithm to the pattern search method by Momma and Bennett (2002), which was designed for the same purpose, revealed that our approach has a better statistical robustness and insensitivity to local optima.

Chapters 5 and 6 dealt with the important problem of ADME *in silico* prediction in modern drug discovery. In Chapter 5 we experimentally demonstrated the usefulness of a SVM-based descriptor selection and proposed the REA algorithm to further improve the descriptor ranking computed by RFE (Guyon et al., 2002). Our experiments showed a clear improvement of the generalization error bounds when compared to the original RFE solution, which suggests a positive effect on the statistical stability by our algorithm. Furthermore, the additional computational demand seems affordable. We applied our algorithm to search for the relevant descriptors in several QSPR and ADME datasets and found our results, if available, to be confirmed in the literature. We also investigated a simple method to incorporate prior knowledge on relevant descriptors into the descriptor selection process. On artificial toy data we could demonstrate the principle usefulness of such an approach.

Chapter 6 introduced an alternative approach for building QSAR/QSPR models by means of kernel functions for molecular graphs. We defined the optimal assignment kernel as a similarity measure for a pair of molecules. It is based on the idea of assigning each atom of the smaller molecule to exactly one from the larger molecule, such that the sum of atom similarity scores is maximized. Thereby the atom similarity score includes information on the physico-chemical properties of each single atom as well as on its neighborhood. By shifting the spectrum of the original similarity matrix on the training data we could cast our similarity measure into a valid symmetric and positive semidefinite Mercer kernel, which can be used in combination with any kernel-based machine learning algorithm. It can be calculated efficiently in cubic time complexity. In practice the average time needed for one kernel evaluation was in most cases below 5ms on one single core AMD Opteron 2.4GHz machine. Experimental comparisons to the marginalized graph kernel by Kashima et al. (2003) in several cases revealed a significant improvement by our approach. At the same time our results were at least as good as those achieved by a classical descriptor model with automatic RFE descriptor selection.

We investigated two major extensions of our method: First, the incorporation of expert provided relevant descriptor information by means of a sum of kernels, multi-kernel SVMs (Weston, 1999), semidefinite programming (Lanckriet et al., 2004) or Kernel CCA (Lai and Fyfe, 2000; Akaho, 2001; Bach and Jordan, 2002). Second, a reduced graph representation of the molecule, in which certain structural features are collapsed into one single node of the molecular graph. We demonstrated that especially the first extension can lead to a further significant reduction of the error rate, whereas the major benefit of the reduced graph representation lies in the fact that expert knowledge on important structural features can be included. In conclusion of our results we speculate that a careful choice of a reduced graph representation together with a conservative selection of descriptors, which are highly relevant, has the highest potential to lead to good QSAR/QSPR models, because this way the most expert knowledge can be incorporated. For the selection of the relevant descriptors the REA algorithm from Chapter 5 might be considered.

In Chapter 7 we investigated the question, which features of the MP in cortical neurons *in vivo* trigger the initiation of an AP. We used a SVM to discriminate between time points in a MP recording prior to an AP and others, which are not close to an AP. For each time point we defined a set of 15 features, which allowed a prediction of the occurrence of an AP with very high fidelity. A further RFE-like feature subset selection over 11 *in vivo* recordings in parallel revealed four basic factors that could explain the AP generation at least as good as all 15 features: the value and the MP rate of change just before AP onset, an increase of the MP over a short period before AP onset, and the average value of the MP over a longer time. Thereby the factor with the highest impact was the MP rate of change just before AP onset. The importance of this feature was also confirmed by the literature (Azouz and Gray, 2000), where a correlation between the potential and the first derivative of the MP at AP onset was proposed. Our approach generalized these results by using a set of 15 features without any a-priori assumptions on their relevance.

Chapter 8 presented kernel-based approaches to cluster genes according to their function solely based on the Gene Ontology (The Gene Ontology Consortium, 2004). A first approach was to define the features for each gene by its minimal GO term distance to all other genes in the dataset. This gave rise to an empirical kernel map (Schölkopf and Smola, 2002). The method was further extended by a PCA preprocessing of the defined feature vector to reduce dimensionality and to achieve a higher robustness. A second approach was to employ the optimal assignment kernel in combination with an appropriately defined kernel function for GO terms. The potential advantage of the second approach is an improved scaling behavior for larger datasets. However, experimental results using dual k-means, spectral clustering and average linkage showed a better clustering behavior of the empirical kernel map, especially when PCA preprocessing was employed. Thereby we achieved the best results using the dual k-means algorithm, whereas average linkage was clearly inferior. Nevertheless, using dual k-means in combination with the empirical kernel map as well as with the optimal assignment kernel we demonstrated that both methods were able to detect meaningful clusters.

We think this work revealed the usefulness of kernel-based pattern analysis in a broad range of practical tasks. Although each individual problem in this thesis required an individual solution, kernel methods have an underlying design principle that can be applied in many domains due to its flexibility and ability to incorporate expert knowledge. However, we do not claim that the kernel-based approach is the only or best way of dealing with the problems presented here. It is just **one** approach, which from our point of view has some nice theoretical advantages that we already pointed out in the introduction.

We also do not claim that our solutions for the individual problems are necessarily the only or best ones. Much more, the goal of this thesis was to point out possible approaches to the given problems using kernel methods. Therefore, our focus was relatively restricted. It thus remains an open issue to eventually find better solutions for the practical problems presented here, maybe following completely different paradigms.

Appendix A

UCI Benchmark Datasets

A.1 Iris Plant Dataset

- **Type:** classification (3 types of iris plant)
- **Features:** 4 (real-valued)
- **Number of instances:** 150
- **Remark:** One class is linearly separable from the others, but the other two are not linearly separable from each other

Table A.1: Number of instances per class

class 1	class 2	class 3
50	50	50

A.2 Glass Identification Dataset

- **Type:** classification (6 types of glass)
- **Features:** 9 (real-valued)
- **Number of instances:** 214

Table A.2: Number of instances per class

class 1	class 2	class 3	class 4	class 5	class 6
70	17	76	13	9	29

A.3 Wine Recognition Dataset

- **Type:** classification (3 types of wine)
- **Features:** 12 (real-valued)
- **Number of instances:** 178

Table A.3: Number of instances per class

class 1	class 2	class 3
59	71	48

A.4 Cleveland Heart Dataset

- **Type:** classification (angiographic heart disease status ranging from 0 to 4)
- **Features:** 13 (5 real-valued, 8 discrete)
- **Number of instances:** 297 from 303 after removing instances with missing values

Table A.4: Number of instances per class

class 1	class 2	class 3	class 4	class 5
160	54	35	35	13

A.5 Wisconsin Breast Cancer

- **Type:** classification (malignant vs. benign)
- **Features:** 9 (discrete)
- **Number of instances:** 683 from 699 after removing instances with missing values

Table A.5: Number of instances per class

class 1 (malginant)	class 2 (benign)
239	444

A.6 Boston Housing

- **Type:** regression (housing prices in suburbs of Boston)
- **Features:** 13 (12 real-valued,1 binary)
- **Number of instances:** 506

A.7 Triazines

- **Type:** regression (inhibition of dihydrofolate reductase by triazines)
- **Features:** 60 (real-valued)
- **Number of instances:** 186, organized for 6-fold cross-validation

A.8 Pyrimidines

- **Type:** regression (inhibition of dihydrofolate reductase by pyrimidines)
- **Features:** 27 (real-valued)
- **Number of instances:** 55, organized for 5-fold cross-validation

A.9 Auto-MPG

- **Type:** regression (city-cycle fuel consumption of cars)
- **Features:** 8 (6 real-valued, 2 discrete)
- **Number of instances:** 398

Appendix B

Test Functions

B.1 Goldstein-Price

- $f(x_1, x_2) = (1 + (x_1 + x_2 + 1)^2(19 - 14x_1 + 3x_2 - 14x_2 + 6x_1x_2 + 3x_2^2)) \cdot (30 + (2x_1 - 3x_2)^2(18 - 32x_1 + 12x_1^2 + 48x_2 - 36x_1x_2 + 27x_2^2))$ $(-2 \leq x_i \leq 2, i = 1, 2)$
- global minimum: $f(x_1^*, x_2^*) = -3$

B.2 Branin

- $f(x_1, x_2) = \left(x_2 - \frac{5x_1^2}{4\pi^2} + \frac{5x_1}{\pi} - 6\right)^2 + 10\left(1 - \frac{1}{8\pi}\right)\cos(x_1) + 10$ $(-5 \leq x_1 \leq 10, 0 \leq x_2 \leq 15)$
- global minimum: $f(x_1^*, x_2^*) \approx -0.3979$

B.3 Michalewicz

- $f(\mathbf{x}) = -\sum_{i=1}^{n}\left(\sin(x_i)\sin^{20}\left(\frac{ix_i^2}{\pi}\right)\right)$ $(0 \leq x_i \leq \pi, i = 1, 2, ..., n)$
- global minimum n = 5: $f(\mathbf{x}) \approx -4.687$

Appendix C

Result Tables Chapter 4

Table C.1: 5-fold cross-validation error $\pm$ standard error on classification (first part of Table) and regression data (second part). For the classification data the mean classification loss in %, for the regression data the mean squared error is reported. Significant wins of EPSGO compared to PS are marked by "*", losses by "-". Significant wins or losses compared to grid search are marked by "**" and "—", respectively.

Data	EPSGO	grid search	PS
2 spirals	0.25 ± 0.30*	0.75 ± 0.50	23.25 ± 9.80
Iris	3.43 ± 1.90*	4.74 ± 1.70	6.87 ± 3.10
Glass	29.15 ± 2.60	31.51 ± 3.30	29.60 ± 1.80
Wine	1.10 ± 0.70*	1.68 ± 1.10	10.27 ± 6.90
Heart	43.78 ± 2.60	43.42 ± 2.40	47.15 ± 2.40
Cancer	3.08 ± 0.30*	3.08 ± 2.70	7.77 ± 4.10
Sinc $(\cdot 10^{-2})$	14.18 ± 2.30*	37.06 ± 2.10	19.51 ± 5.50
Housing $(\cdot 10^{-2})$	15.35 ± 5.30*	13.51 ± 2.20	16.98 ± 2.80
Triazines$(\cdot 10^{-4})$	6.95 ± 3.40*	7.29 ± 3.70	25.00 ± 16.00
Pyrimidines $(\cdot 10^{-2})$	0.20 ± 0.10*	0.23 ± 0.20	53.68 ± 13.60
Auto-MPG $(\cdot 10^{-2})$	11.79 ± 1.70*	12.49 ± 1.90	19.51 ± 5.50

Table C.2: Comparison of EPSGO, EPSGO-Bag and PS-Bag. 5-fold cross-validation error $\pm$ standard error on classification (first part of Table) and regression data (second part). Significant wins of EPSGO/EPSGO-Bag compared to PS-Bag are marked by "*", losses by "-". Significant wins of EPSGO-Bag compared to EPSGO are marked by "**", losses by "—".

Data	**EPSGO**	**EPSGO-Bag**	**PS-Bag**
2 spirals	$0.25 \pm 0.30^*$	$1.25 \pm 0.70^*$	8.75 ± 2.90
Iris	$3.43 \pm 1.90^-$	$0.00 \pm 0.00^{**}$	0.00 ± 0.00
Glass	$29.15 \pm 2.60^-$	$2.75 \pm 3.00^{**}$	0.00 ± 0.00
Wine	$1.10 \pm 0.70^-$	0.00 ± 0.00	0.00 ± 0.00
Heart	$43.78 \pm 2.60^-$	$2.61 \pm 2.60^{**}$	4.29 ± 4.30
Cancer	3.08 ± 0.30	3.37 ± 0.60	3.66 ± 0.70
Sinc $(\cdot 10^{-2})$	$14.18 \pm 2.30^*$	$15.15 \pm 2.20^*$	71.05 ± 21.70
Housing $(\cdot 10^{-2})$	$15.35 \pm 5.30^*$	$14.32 \pm 3.20^*$	28.60 ± 3.80
Triazines $(\cdot 10^{-4})$	$6.95 \pm 3.40^*$	$6.24 \pm 3.60^*$	59.00 ± 13.00
Pyrimidines $(\cdot 10^{-2})$	$0.20 \pm 0.10^*$	$0.59 \pm 0.40^*$	1.50 ± 0.70
Auto-MPG $(\cdot 10^{-2})$	$11.79 \pm 1.70^*$	$12.84 \pm 1.80^*$	19.09 ± 2.20

Appendix D

Result Tables Chapter 5

Table D.1: 10-fold CV $\pm$ standard error. For classification problems (first part of Table) the class loss (%) is reported, for the regression data (second part) the mean squared error (MSE) $\times 100$. Significant improvements compared to the SVM model are marked by "*", losses by "-". Likewise, significant wins compared to the FCBF algorithm are denoted by "**" and "–", respectively.

Method	**HIA**	**Yoshida**	**SOL**	**BBB**	**Huuskonen**
SVM/SVR	19.6 ± 3.72	33.93 ± 3.72	7.77 ± 1.42	76.03 ± 10.24	15.29 ± 0.81
EXP.	15.33 ± 2.55	38.13 ± 1.88	–	$\mathbf{38.10 \pm 5.38}^{*,**}$	48.96 ± 3.21
FCBF	22.02 ± 3.38	32.82 ± 2.01	$15.54 \pm 1.13^{-}$	80.19 ± 15.05	$55.32 \pm 3.27^{-}$
RFE	18.42 ± 3.36	$\mathbf{32.38 \pm 3.61}$	$\mathbf{2.70 \pm 1.40}^{*,**}$	$54.1 \pm 9.46^{*}$	$\mathbf{10.68 \pm 0.96}^{*,**}$
REA	20.22 ± 3.36	33.95 ± 2.87	$3.37 \pm 1.65^{*,**}$	$53.82 \pm 10.54^{*}$	$11.05 \pm 0.97^{*,**}$
RFE+EXP.	18.42 ± 2.7	33.56 ± 3.14	–	59.41 ± 11.80	$\mathbf{10.68 \pm 0.96}^{*,**}$
REA+EXP.	16.58 ± 2.68	34.34 ± 3.30	–	$56.54 \pm 10.23^{*}$	$11.05 \pm 0.97^{*,**}$
CART	15.99 ± 2.72	38.11 ± 2.64	10.81 ± 1.99	70.79 ± 13.41	$21.24 \pm 0.93^{-,**}$

Table D.2: 10-fold CV ± standard error for the regression datasets. Reported is the squared correlation (%). Significant improvements compared to the SVM model are marked by "*", losses by "-". Likewise, significant wins compared to the FCBF algorithm are denoted by "**" and "–", respectively.

Method	BBB	Huuskonen
SVM/SVR	36.02 ± 7.54	84.79 ± 0.70
EXP.	$\mathbf{65.28 \pm 6.08}^{*,**}$	$56.65 \pm 2.03^{-}$
FCBF	32.30 ± 5.41	$48.74 \pm 3.44^{-}$
RFE	49.89 ± 9.53	$\mathbf{89.98 \pm 0.37}^{*,**}$
REA	$50.01 \pm 8.59^{*}$	$89.32 \pm 0.66^{*,**}$
RFE+EXP.	$52.70 \pm 8.74^{*,**}$	$89.61 \pm 0.71^{*,**}$
REA+EXP.	$56.33 \pm 9.81^{*,**}$	$89.32 \pm 0.66^{*,**}$
CART	$53.60 \pm 7.63^{*,**}$	$79.22 \pm 1.59^{-,**}$

Table D.3: REA algorithm: average improvement (%) ± standard deviation on generalization error bound compared to the original ranking (first row) and number of loops needed until termination (second row).

Method	HIA	Yoshida	SOL	BBB	Huuskonen
REA	28.71 ± 19.56 14 ± 5	8.53 ± 7.68 13 ± 4	12.53 ± 16.61 16 ± 3	42.53 ± 15.99 27 ± 18	18.11 ± 6.1 30 ± 7
REA+EXP.	30.56 ± 19.68 15 ± 2	9.38 ± 6.72 13 ± 4	–	18.07 ± 17.04 16 ± 6	18.11 ± 6.1 30 ± 7

Table D.4: Descriptors selected by the REA method on the Huuskonen dataset during all training runs.

Descriptor name
logP
polar surface area (PSA)
Burden's modified eigenvalues: electronegativity
Burden's modified eigenvalues: atom in ACC
Burden's modified eigenvalues: atom in DON/ACC
Burden's modified eigenvalues: atom is positive
Burden's modified eigenvalues: partial charge
Burden's modified eigenvalues: implicit valence
geometrical radius
#H-bond donors
#H-bond acceptors
#basic groups
#S-atoms
#N-atoms

Appendix E

Result Tables Chapter 6

Table E.1: Atom and bond features chosen for the optimal assignment kernel

features	nominal	real valued
atom	in donor, in acceptor, in donor or acceptor (Böhm and Klebe, 2002), in terminal carbon, in aromatic system (Bonchev and Rouvray, 1990), negative/positive, in ring (Figueras, 1996), in conjugated environment, free electrons, implicit valence, heavy valence, hybridization, is chiral, is axial	electrotopological state, mass, graph potentials, electron-affinity, van-der-Waals volume, electrogeometrical state, electronegativity (Pauling), intrinsic state (Todeschini and Consonni, 2000), Gasteiger/Marsili partial charge (Gasteiger and Marsili, 1978)
bond	order, in aromatic system(Bonchev and Rouvray, 1990), in ring (Figueras, 1996), is rotor, in carbonyl/amide/ primary amide/ester group	geometric length

Table E.2: 10-fold CV error $\pm$ std. error. For classification problems (first part of Table) the class loss (%) is reported, for the regression data (second part) the mean squared error (MSE) $\times 100$. Significant wins of the the OA/OARG kernel compared to the MG kernel and the DESC/DESCSEL model are denoted by "*" and "**", losses by "-" and "–", respectively.

Method	HIA	Yoshida	SOL	BBB	Huuskonen
DESC	19.60 ± 3.72	33.93 ± 3.72	7.77 ± 1.47	76.03 ± 10.24	15.29 ± 0.81
DESCSEL	18.42 ± 3.26	32.38 ± 3.61	2.36 ± 1.32	54.10 ± 9.46	$\mathbf{10.68 \pm 0.96}$
MG	17.13 ± 3.14	33.63 ± 1.99	1.34 ± 0.74	52.17 ± 6.09	12.16 ± 0.67
OA	15.26 ± 2.74	$30.98 \pm 2.07^{*}$	$\mathbf{1.01 \pm 0.52}^{**}$	$\mathbf{36.24 \pm 5.48}^{*,**}$	$11.17 \pm 0.6^{*,**}$
OARG	$\mathbf{14.04 \pm 2.22}^{*}$	$\mathbf{29.89 \pm 1.86}^{*}$	$1.37 \pm 0.56^{**}$	$37.47 \pm 4.34^{*,**}$	$21.1 \pm 1.23^{-,--}$

Table E.3: 10-fold CV error $\pm$ std. error. for the regression datasets. Reported is the squared correlation (%). Significant wins of the the OA/OARG kernel compared to the MG kernel and the DESC/DESCSEL model are denoted by "*" and "**", losses by "-" and "/", respectively.

Method	BBB	Huuskonen
DESC	36.02 ± 7.54	84.79 ± 0.71
DESCSEL	49.89 ± 9.53	$\mathbf{89.61 \pm 0.71}$
MG	57.56 ± 4.76	87.93 ± 0.53
OA	$\mathbf{63.75 \pm 6.02}^{**}$	$88.98 \pm 0.37^{*,**}$
OARG	$60.28 \pm 6.04^{**}$	$79.37 \pm 0.75^{-,--}$

Table E.4: Effect of incorporating expert knowledge. Significant improvements compared to the original OA/OARG kernels are marked by "*", losses by "-". Likewise, significant wins/losses compared to the EXP model are marked by "**" and "–".

Method	HIA	Yoshida	BBB	Huuskonen
EXP	15.33 ± 2.55	38.13 ± 1.88	38.10 ± 5.38	48.96 ± 3.21
OA+EXP	$\mathbf{12.13 \pm 2.54}$	35.81 ± 1.59	36.99 ± 5.13	$\mathbf{10.95 \pm 0.57}^{**}$
OARG+EXP	12.17 ± 2.89	32.38 ± 2.65	30.79 ± 4.87	$13.84 \pm 0.78^{*,**}$
OA+MKSVM	14.60 ± 3.18	34.63 ± 2.65	32.70 ± 4.16	—
OARG+MKSVM	13.35 ± 3.56	$31.67 \pm 2.32^{**}$	$30.81 \pm 3.12^{*}$	—
OA+SDP	14.59 ± 1.83	$30.54 \pm 2.09^{**}$	32.95 ± 4.80	—
OARG+SDP	14.60 ± 1.83	$\mathbf{28.26 \pm 2.38}^{**}$	$\mathbf{30.31 \pm 4.12}^{*}$	—
OA+KCCA	16.99 ± 3.24	36.58 ± 2.62	35.48 ± 4.78	$20.05 \pm 0.92^{-,**}$
OARG+KCCA	15.85 ± 2.91	34.34 ± 2.54	36.58 ± 5.11	$20.97 \pm 0.92^{**}$

Table E.5: Effect of incorporating expert knowledge (regression datasets). Reported is the squared correlation (%). Significant improvements compared to the original OA/OARG kernels are marked by "*", losses by "-". Likewise, significant wins/losses compared to the EXP model are marked by "**" and "–".

Method	BBB	Huuskonen
EXP	65.28 ± 6.08	56.65 ± 2.03
OA+EXP	60.86 ± 6.26	$\mathbf{89.10 \pm 0.43}^{*,**}$
OARG+EXP	66.44 ± 4.75	$86.26 \pm 0.65^{*,**}$
OA+MKSVM	64.92 ± 5.39	—
OARG+MKSVM	$66.28 \pm 4.75^{*}$	—
OA+SDP	63.70 ± 6.72	—
OARG+SDP	65.31 ± 4.12	—
OA+KCCA	$\mathbf{66.56 \pm 4.33}$	$79.95 \pm 0.96^{-,**}$
OARG+KCCA	63.26 ± 5.91	$79.07 \pm 0.90^{**}$

Table E.6: SMARTS patterns used for defining the reduced graph representation.

Name	SMARTS
ring	[R]
H-bond donor	[$([NH2]-a), ND1H3, ND2H2, ND3H1, ND2H1, nD1H3, nD2H2, nD3H1, nD2H1, $(Cl-*), $(Br-*), $(I-*)]
H-bond acceptor	[$(N#C-[C,c]), OD1X1, oD1X1, OD2X2, oD2X2, ND3X3, nD3X2, ND2X2, nD2X2, ND1X1, nD1X1]
H-bond donor/acceptor	[$([NH2]-A), $([OH]-*)]
terminal carbon	([CH3, CD1H2, CD1H1, cH3, cD1H2, cD1H1]
positive charge	[+,++,+++]
negative charge	[-,–,—]

Appendix F

Result Tables Chapter 8

Table F.1: A cluster with genes related to metabolism (first dataset) found identically by k-means + EKM with and without PCA preprocessing.

Acc. number	Gene Ontology terms
AA001722	ATP catabolism citrate metabolism coenzyme A metabolism lipid metabolism
AA011388	biotin metabolism fatty acid biosynthesis
AA025800	L-serine biosynthesis
AA026314	tetrahydrobiopterin biosynthesis
AA039466	fatty acid biosynthesis
AA040861	UDP-N-acetylglucosamine biosynthesis
AA043362	circulation fatty acid metabolism lipid catabolism lipid transport posttranslational membrane targeting
AA043796	lactose biosynthesis
AA044444	glycolysis
AA045181	C21-steroid hormone biosynthesis cholesterol metabolism lipid metabolism mitochondrial transport steroid metabolism

continue next page

Table F.1 – continued

AA045372	cholesterol biosynthesis isoprenoid biosynthesis steroid biosynthesis
AA053173	cholesterol biosynthesis steroid biosynthesis
AA053331	cholesterol biosynthesis
AA053461	asparagine biosynthesis glutamine metabolism
AA054956	fatty acid biosynthesis
AA057761	glycolysis
AA053028	cholesterol biosynthesis cholesterol metabolism germ-cell migration gonad development
H12318	fatty acid beta-oxidation fatty acid metabolism
H63779	central nervous system development epidermal differentiation lipid metabolism peripheral nervous system development
H70783	fatty acid biosynthesis
N32784	neurotransmitter biosynthesis and storage nitric oxide biosynthesis phenylalanine catabolism
N35315	amino acid metabolism
N91268	lipid metabolism steroid biosynthesis
R00824	L-serine biosynthesis L-serine metabolism
R12563	fatty acid desaturation
R60996	aerobic respiration electron transport tricarboxylic acid cycle
T40987	electron transport
W88807	uroporphyrinogen III biosynthesis
W89012	fatty acid beta-oxidation fatty acid metabolism

continue next page

Table F.1 – continued

W91979	cholesterol biosynthesis

Table F.2: A cluster with genes related to metabolusm and RNA processing (first dataset) found by k-means + OA-Term.

Acc. number	Gene Ontology terms
AA010407	**RNA processing**
AA011388	biotin metabolism fatty acid biosynthesis
AA025800	L-serine biosynthesis
AA026314	tetrahydrobiopterin biosynthesis
AA035360	regulation of transcription, DNA-dependent
AA039466	fatty acid biosynthesis
AA045181	C21-steroid hormone biosynthesis cholesterol metabolism lipid metabolism mitochondrial transport steroid metabolism
AA045372	cholesterol biosynthesis isoprenoid biosynthesis steroid biosynthesis
AA053173	cholesterol biosynthesis steroid biosynthesis
AA053331	cholesterol biosynthesis
AA053461	asparagine biosynthesis glutamine metabolism
AA054956	fatty acid biosynthesis
AA055585	regulation of transcription, DNA-dependent
AA056338	ribosome biogenesis **rRNA processing**
H12318	fatty acid beta-oxidation fatty acid metabolism
H27557	regulation of transcription, DNA-dependent
H70783	fatty acid biosynthesis
N35315	amino acid metabolism
N47794	**RNA catabolism**
N91268	lipid metabolism steroid biosynthesis
R00824	L-serine biosynthesis L-serine metabolism
R09377	**RNA processing**

continue next page

Table F.2 – continued

R12563	fatty acid desaturation
R39209	regulation of transcription, DNA-dependent
R43728	**rRNA processing**
T50056	regulation of transcription, DNA-dependent
R49309	regulation of transcription, DNA-dependent
W44416	drug resistance glutamine metabolism **nucleobase, nucleoside, nucleotide and nucleic acid metabolism** 'de novo' pyrimidine base biosynthesis
W70150	regulation of transcription, DNA-dependent
W88807	uroporphyrinogen III biosynthesis
W89012	fatty acid beta-oxidation fatty acid metabolism
W91979	cholesterol biosynthesis

Table F.3: A cluster with genes related to DNA replication (second dataset) found by k-means + OA-Term.

Acc. number	Gene Ontology terms
D16562_at	**energy pathways**
D26018_at	DNA dependent DNA replication
D26535_at	**energy pathways**
D38073_at	DNA replication initiation
D50370_at	nucleosome assembly
D79984_at	chromatin assembly/disassembly *regulation of transcription from Pol II promoter*
L07541_at	DNA strand elongation
M87339_at	DNA strand elongation
U20980_at	DNA replication dependent nucleosome assembly protein complex assembly
U28413_at	DNA repair
U28749_at	establishment and/or maintenance of chromatin architecture *regulation of transcription, DNA-dependent* development
X55740_at	DNA metabolism
X62153_at	DNA replication initiation
X74331_at	DNA replication, priming
Y00764_at	**electron transport** **oxidative phosphorylation** **aerobic respiration**

Table F.4: A cluster with genes related to DNA replication (second dataset) found by k-means + EKM + PCA.

Acc. number	Gene Ontology terms
D26018_at	DNA dependent DNA replication
D38073_at	DNA replication initiation
D38551_at	double-strand break repair DNA recombination meiotic recombination
D50370_at	nucleosome assembly
J04611_at	DNA ligation double-strand break repair double-strand break repair via nonhomologous end-joining DNA recombination
L07541_at	DNA strand elongation
M60974_at	regulation of cell cycle regulation of CDK activity DNA repair apoptosis response to stress cell cycle arrest
M87339_at	DNA strand elongation
U27516_at	double-strand break repair mitotic recombination meiotic recombination
X62153_at	DNA replication initiation
X74331_at	DNA replication, priming

Table F.5: A cluster with genes related to DNA replication (second dataset) found by k-means + EKM.

Acc. number	Gene Ontology terms
D26018_at	DNA dependent DNA replication
D38073_at	DNA replication initiation
D38551_at	double-strand break repair DNA recombination meiotic recombination
D50370_at	nucleosome assembly
J04611_at	DNA ligation double-strand break repair double-strand break repair via nonhomologous end-joining DNA recombination
L07541_at	DNA strand elongation
M87339_at	DNA strand elongation
U27516_at	double-strand break repair mitotic recombination meiotic recombination
X62153_at	DNA replication initiation
X67155_at	mitotic spindle elongation
X74331_at	DNA replication, priming

Bibliography

Abolmaali, S., Ostermann, C., and Zell, A. (2003). The compressed feature matrix - a novel descriptor for adaptive similarity search. *J. Mol. Model.*, 9:66 – 75.

Adryan, B. and Schuh, R. (2004). Gene Ontology-based clustering of gene expression data. *To appear in Bioinformatics*.

Agrafiotis, D. and Xu, H. (2003). A Geodisc Framework for Analyzing Molecular Similarities. *J. Chem. Inf. Comp. Sci.*, 43:475 – 484.

Aizerman, M., Braverman, E., and Rozonoer, L. (1964). Theoretical foundations of the potential function method in pattern recognition learning. *Automation and Remote Control*, 25:821 – 837.

Akaho, S. (2001). A kernel method for canonical correlation analysis. In *Proc. Int. Meeting Psychometric Society*.

Aronszajn, N. (1950). Theory of reproducing kernels. *Trans. Am. Math. Soc.*, 68:337 – 404.

Artursson, P. and Bergström, C. (2003). Intestinal absorption: The role of polar surface area. In van de Waterbeemd, H., Lennernäs, H., and Artursson, P., editors, *Drug Bioavailability*, pages 341 – 357. Wiley-VCH, Weinheim.

Azouz, R. and Gray, C. (2000). Dynamic spike threshold reveals a mechanism for synaptic coincidence detection in cortical neurons in vivo. *Proc. Nat. Acad. Sci.*, 97:8110 – 8115.

Bach, F. and Jordan, M. (2002). Kernel independent component analysis. *J. Machine Learning Research*, 3:1 – 48.

Balon, K., Riebesehl, B., and Müller, B. (1999). Drug liposome partitioning as a tool for the prediction of human passive intestinal absorption. *Pharm. Res.*, 16:882 – 888.

Bartlett, P. and Shawe-Taylor, J. (1999). Generalization performance of support vector machines and other pattern classifiers. In Schölkopf, B., Burges, C., and Smola, A., editors, *Advances in Kernel Methods - Support Vector Learning*, pages 43 – 54. MIT Press, Cambridge, MA.

Battiti, R. (1994). Using mutual information for selecting features in supervised neural net learning. *IEEE Trans. Neural Networks*, 5(4):537 – 550.

Beißbarth, T. and Speed, T. (2004). GOstat: finding statistically overexpressed Gene Ontologies within groups of genes. *Bioinformatics*, 20(9):1464–1465.

Ben-Hur, A. and Noble, W. (2005). Kernel Methods for Predicting Protein-Protein Interactions. *Bioinformatics*, 21(1):i38 – i46.

Bender, A. and Glen, R. (2004). Molecular similarity: a key technique in molecular informatics. *Org. Biomol. Chem.*, 2:3204 – 3218.

Bishop, C. (1995). *Neural Networks for Pattern Recognition*. Clarendon Press, Oxford.

Blum, A. L. and Langley, P. (1997). Selection of Relevant Features and Examples in Machine Learning. *Artificial Intelligence*, 97(12):245 – 271.

Böhm, H.-J. and Schneider, G. (2000). *Virtual Screening for Bioactive Molecules*. Wiley-VCH, Weinheim.

Böhm, M. and Klebe, G. (2002). Development of a new hydrogen-bond descriptor and their application to comparative mean field analysis. *J. Med. Chem.*, 45:1585 – 1597.

Bonchev, D. and Rouvray, D. H., editors (1990). *Chemical Graph Theory: Introduction and Fundamentals*, volume 1 of *Mathematical Chemistry Series*. Gordon and Breach Science Publishers, London, UK.

Boser, B., Guyon, M., and Vapnik, V. (1992). A training algorithm for optimal margin classifiers. In Haussler, D., editor, *Proc. 5th Ann. ACM Workshop on Comp. Learning Theory*, Pittsburgh, PA. ACM Press.

Bradley, P. and Mangasarian, O. (1998). Feature selection via concave minimization and support vector machines. In *Proc. 13th Int. Conf. Machine Learning*, pages 82 – 90.

Breiman, L. (2001). Random forests. *Machine Learning*, 45(1):5 – 32.

Breiman, L., Friedman, J., Olshen, R., and Stone, C. (1984). *Classification and Regression Trees*. Wadsworth and Brooks.

Budanitsky, A. and Hirst, G. (2001). Semantic distance in WordNet: An experimental, application-oriented evaluation of five measures. In *Workshop on WordNet and other Lexical Resources, Second meeting of the Nord American Chapter of the Association for Computational Linguistics*. Pittsburgh.

Byvatov, E., Fechner, U., Sadowski, J., and Schneider, G. (2003). Comparison of Support Vector Machine and Artificial Neural Network Systems for Drug/Nondrug Classification. *J. Chem. Inf. Comput. Sci.*, 43(6):1882 – 1889.

Byvatov, E. and Schneider, G. (2004). Svm-based feature selection for characterization of focused compound collections. *J. Chem. Inf. Comp. Sci.*, 44(3):993 – 999.

Carhart, R., Smith, D., and Venkataraghavan, R. (1985). Atom pairs as molecular features in structure-activity studies: Definitionand applications. *J. Chem. Inf. Comput. Sci.*, 25:64 – 73.

Chang, C. and Lin, C. (2001). *LIBSVM: a library for support vector machines*. Available from http://www.csie.ntu.edu.tw/~cjlin/libsvm.

Chapelle, O. and Vapnik, V. (2000). Model selection for Support Vector Machines. In Solla, S., Leen, T., and Müller, K.-R., editors, *Adv. Neural Inf. Proc. Syst. 12*, Cambridge, MA. MIT Press.

Chapelle, O., Vapnik, V., Bousqet, O., and Mukherjee, S. (2002). Choosing Multiple Parameters for Support Vector Machines. *Machine Learning*, 46(1):131 – 159.

Chen, X., Rusinko, A., Tropsha, A., and Young, S. S. (1999). Automated Pharmacophore Identification for Large Chemical Data Sets. *J. Chem. Inf. Comput. Sci.*, 39:887–896.

Cherkassky, V. and Ma, Y. (2004). Practical selection of svm parameters and noise estimation for svm regression. *Neural Networks*, 17(1):113 – 126.

Cho, R., Huang, M., Campbell, M., Dong, H., Steinmetz, L., Sapinoso, L., Hampton, G., Elledge, S., Davis, R., and Lockhart, D. (2001). Transcriptional regulation and function during the human cell cycle. *Nature Genetics*, 27(1):48–54.

Chung, F. (1997). *Spectral Graph Theory*. Number 92 in CBMS Regional Conference Series in Mathematics. American Mathematical Society, Providence, RI.

Cortes, C. and Vapnik, V. (1995). Support vector networks. *Machine Learning*, 20:273 – 297.

Cramer, R., Patterson, D., and Bunce, J. (1988). Comparative molecular field analysis (comfa). 1. effect of shape on binding of steroids to carrier proteins. *J. Am. Chem. Soc.*, 110:5959 – 5967.

Cristianini, N. and Shawe-Taylor, J. (2000). *An Introduction to Support Vector Machines*. Cambridge University Press.

Csato, L. and Opper, M. (2002). Sparse online gaussian processes. *Neural Computation*, 14(3):641 – 669.

Davies, S. and Russel, S. (1994). NP-Completeness of Searches for Smallest Possible Feature Sets. In *Proc. 1994 AAAI Fall Symposion on Relevance*, pages 37 – 39.

Dempster, A., Laird, N., and Rubin, D. (1977). Maximum likelihood from incomplete data via the em algorithm. *J. Royal Statistical Soc., Series B*, 39(1):1 – 38.

Dennis, J. and Torczon, V. (1994). Derivative-free pattern search methods for multidisciplinary design problems. In *Proc. 5th AIAA/USAF/NASA/ISSMO Symposium on Multidisciplinary Analysis and Optimization*, pages 922 – 932.

Destexhe, A. and Pare, D. (1999). Impact of network activity on the integrative properties of neocortical pyramidal neurons in vivo. *J. Neurophysiol.*, 81:1531 – 1547.

Ding, C., He, X., Zha, H., Gu, M., and Simon, H. (2001). Spectral min-max cut for graph partitioning and data clustering. In *Proc. 1st IEEE Int. Conf. Data Mining*, pages 107 –114.

Doniger, S., N.Salomonis, Dahlqusi, K., Vranizan, K., Lawlor, S., and Conklin, B. (2003). MAPPFinder: using Gene Ontology and GenMAPP to create a global gene-expression profile from microarray data. *Genome Biology*, 4(1):R7.

Duda, R., Hart, P., and Stork, D. (2001). *Pattern Classification*. Wiley-Interscience, New York.

Eisen, M., Spellman, P., Botstein, D., and Brown, P. (1998). Cluster analysis and display of genome-wide expression patterns. In *Proceedings of the National Academy of Sciences, USA*, volume 95, pages 14863–14867.

Figueras, J. (1996). Ring Perception Using Breadth–First Search. *J. Chem. Inf. Comput. Sci.*, 36:986–991.

Fourcaud-Trocme, N., Hansel, D., van Vreeswijk, C., and Brunel, N. (2000). How spike generation mechanisms determine the neuronal response to fluctuating inputs. *J. Neurosci.*, 23:11628 – 11640.

Freund, Y. and Shapire, R. (1996). A decision-theoretic generalization of on-line learning and an application to boosting. *J. Comp. and Syst. Sci.*, 55(1):119 – 139.

Fröhlich, H., Chapelle, O., and Schölkopf, B. (2004a). Feature Selection for Support Vector Machines using Genetic Algorithms. *Int. J. AI Tools: Special Issue on Selected Papers from the 15th IEEE Int. Conf. on Tools with AI 2003*, 13(4):791 – 800.

Fröhlich, H., Naundorf, B., Volgushev, M., and Wolf, F. (2005a). Which features trigger action potentials in cortical neurons in vivo? In *Proc. Int. Joint Conf. Neural Networks*, pages 250 – 255.

Fröhlich, H., Speer, N., and Zell, A. (2006a). Kernel based functional gene grouping. In *Proc. Int. Joint Conf. Neural Networks*, pages 6886 – 6891.

Fröhlich, H., Wegner, J., Sieker, F., and Zell, A. (2005b). Optimal assignment kernels for attributed molecular graphs. In Raedt, L. D. and Wrobel, S., editors, *Proc. Int. Conf. Machine Learning*, pages 225 – 232. ACM Press.

Fröhlich, H., Wegner, J., Sieker, F., and Zell, A. (2006b). Kernel functions for attributed molecular graphs – a new similarity based approach to adme prediction in classification and regression. *QSAR & Comb. Sci.*, 25(4):317 – 326.

Fröhlich, H., Wegner, J., and Zell, A. (2005c). Assignment kernels for chemical compounds. In *Proc. Int. Joint Conf. Neural Networks*, pages 913 – 918.

Fröhlich, H., Wegner, J. K., and Zell, A. (2004b). Towards Optimal Descriptor Subset Selection with Support Vector Machines in Classification and Regression. *QSAR & Comb. Sci.*, 23:311–318.

Fröhlich, H. and Zell, A. (2004). Feature Subset Selection for Support Vector Machines by Incremental Regularized Risk Minimization. In *Proc. IEEE Int. Joint Conf. on Neural Networks (IJCNN)*, volume 3, pages 2041 – 2046.

Fröhlich, H. and Zell, A. (2005). Efficient parameter selection for support vector machines in classification and regression via model-based global optimization. In *Proc. Int. Joint Conf. Neural Networks*, pages 1431 – 1438.

Gärtner, T., Flach, P., and Wrobel, S. (2003). On graph kernels: Hardness results and efficient alternatives. In *Proc. 16th Ann. Conf. Comp. Learning Theory and 7th Ann. Workshop on Kernel Machines*.

Gasteiger, J. and Marsili, M. (1978). A New Model for Calculating Atomic Charges in Molecules. *Tetrahedron Lett.*, 34:3181–3184.

Gat-Viks, I., Sharan, R., and Shamir, R. (2003). Scoring clustering solutions by their biological relevance. *Bioinformatics*, 19(18):2381–2389.

Gelman, A., Carlin, J., Stern, H., and Rubin, D. (2004). *Bayesian Data Analysis*. Chapman & Hall/CRC.

Gene Lynx (2004). http://www.genelynx.org.

Godden, J., Furr, J., Xue, L., Stahura, F., and Bajorath, J. (2003). Recursive median partitioning for virtual screening of large databases. *J. Chem. Inf. Comp. Sci.*, 43:182 – 188.

Gohlke, H., Dullweber, F., Kamm, W., März, J., Kissel, T., and Klebe, G. (2001). Prediction of human intestinal absorption using a combined 'simmulated annealing/backpropagation neural network' approach. In Hültje, H.-D. and Sippl, W., editors, *Rational Approaches Drug Des.*, pages 261 – 270, Barcelona. Prous Science Press.

Guyon, I. and Elisseeff, A. (2003). An Introduction into Variable and Feature Selection. *J. Machine Learning Research Special Issue on Variable and Feature Selection*, 3:1157 – 1182.

Guyon, I., Weston, J., Barnhill, S., and Vapnik, V. (2002). Gene Selection for Cancer Classification using Support Vector Machines. *Machine Learning*, 46:389 – 422.

Halgren, T. A. (1998). Merck molecular force field. I–V. MMFF94 Basics and Parameters. *J. Comput. Chem.*, 17:490–641.

Hastie, T., Rosset, S., Tishbirani, R., and Zhu, J. (2004). The entire regularization path for support vector machines. *J. Machine Learning Research*, 5:1391 – 1415.

Hastie, T., Tibshirani, R., and Friedman, J. (2001). *The Elements of Statistical Learning*. Springer.

Haussler, D. (1999). Convolution kernels on discrete structures. Technical Report UCSC-CRL-99-10, University of California Santa Cruz.

Heckerman, D. (1997). A tutorial on learning with bayesian networks. *Data Mining and Knowledge Discovery*, 1:79 – 119.

Helma, C., King, R., and Kramer, S. (2001). The predictive toxicology challenge 2000-2001. *Bioinformatics*, 17:107 – 108.

Hodgkin, A. and Huxley, A. (1952). A quantitative description of ion currents and its applications to conduction and excitation in nerve membranes. *J. Physiology*, 117:500 – 544.

Hoeffding, W. (1963). Probability inequalities for sums of bounded random variables. *J. American Statist. Assoc.*, 58:13 – 30.

Hopfield, J. (1982). Neural networks and physical systems with emergent collective computational abilities. *Proc. Nat. Acad. Sci.*, 79:2554 – 2558.

Hotelling, H. (1933). Analysis of a complex of statistical variables into principal components. *J. Educat. Psychol.*, 24:417 – 441 and 498 – 520.

Hotelling, H. (1936). Relations between two sets of variables. *Biometrika*, 28:321 – 377.

Hou, T. and Xu, X. (2003). Adme evaluation in drug discovery. 3. modelling blood-brain barrier partitioning using simple molecular descriptors. *J. Chem. Inf. Comput. Sci.*, 43(6):2137 – 2152.

Huuskonen, J. (2000). Estimation of Aqueous Solubility for Diverse Set of Organic Compounds Based on Molecular Topology. *J. Chem. Inf. Comput. Sci.*, 40:773–777.

Hvidsten, T., Laegreid, A., and Komorowski, J. (2003). Learning rule-based models of biological process from gene expression time profiles using Gene Ontology. *Bioinformatics*, 19(9):1116–1123.

Iyer, V., Eisen, M., Ross, D., Schuler, G., Moore, T., Lee, J., Trent, J., Staudt, L., Jr, J. H., Boguski, M., Lashkari, D., Shalon, D., Botstein, D., and Brown, P. (1999). The transcriptional program in response of human fibroblasts to serum. *Science*, 283:83–87.

Jain, A. and Dubes, R. (1988). *Algorithms for Clustering Data*. Prentice-Hall, Englewood Cliffs, NJ.

Jain, A. K., Murty, M. N., and Flynn, P. J. (1999). Data clustering: a review. *ACM Computing Surveys*, 31(3):264–323.

Jiang, J. and Conrath, D. (1998). Semantic similarity based on corpus statistics and lexical taxonomy. In *Proceedings of the International Conference on Research in Computational Linguistics*, Taiwan.

Joachims, T. (2000). Estimating the generalization performance of a svm efficiently. In *Proc. Int. Conf. Machine Learning*. Morgan Kaufmann.

John, G., Kohavi, R., and Pfleger, K. (1994). Irrelevant features and the subset selection problem. In *Machine Learning: Proc. 11th Int. Conf.*, pages 121 – 129. Morgan Kaufmann.

Jones, D., Perttunnen, C., and Stuckman, B. (1993). Lipschitzian optimization without the lipschitz constant. *J. Optimization Theory and Applications*, 79(1):157 – 181.

Jones, D., Schonlau, M., and Welch, W. (1998). Efficient global optimization of expensive black-box functions. *J. Global Optimization*, 13:455 – 492.

Joslyn, C., Mniszewski, S., Fulmer, A., and Heaton, G. (2004). The gene ontology categorizer. *Bioinformatics*, 20(1):i169–i177.

Kannan, R. and Vempala, S. (2000). On clusterings - good, bad and spectral. In *Proc. Symp. Found. Comp. Sci.*, pages 367 – 377.

Kansy, M., Senner, F., and Gubernator, K. (1998). Physicochemical high throughput screening: Parallel artificial membrane permeation assay in the description of passive absorption processes. *J. Med. Chem.*, 41:1007 – 1010.

Kashima, H., Tsuda, K., and Inokuchi, A. (2003). Marginalized kernels between labeled graphs. In *Proc. 20th Int. Conf. on Machine Learning*.

Kashima, H., Tsuda, K., and Inokuchi, A. (2004). Kernels for graphs. In Schölkopf, B., Tsuda, K., and Vert, J.-P., editors, *Kernel Methods in Computational Biology*, pages 155 – 170. MIT Press, Cambridge, MA.

Kimeldorf, G. and Wahba, G. (1971). Some results on tchebycheffian spline functions. *J. Math. Anal. and Appl.*, 33:82 – 95.

Koch, C., Bernander, O., and Douglas, R. (1995). Do neurons have a voltage or a current threshold for action potential initiation? *Comp. Neuroscience*, 2:63 – 82.

Kohavi, R. and John, G. (1997). Wrappers for Feature Subset Selection. *Artificial Intelligence*, 97(12):273 – 324.

Kohonen, T. (1995). *Self Organizing Maps*. Springer, Berlin.

Krishnapuram, B., Carin, L., and Hartemink, A. (2003). Joint classifier and feature optimization for cancer diagnosis using gene expression data. In *Proc. An. Int. Conf. Research in computational molecular biology*, pages 167 – 175. ACM Press, NY.

Kubinyi, H. (2002). From Narcosis to Hyperspace: The History of QSAR. *Quant. Struct. Act. Relat.*, 21:348–356.

Kubinyi, H. (2003). Drug research: myths, hype and reality. *Nature Reviews: Drug Discovery*, 2:665–668.

Kubinyi, H. (2004). Changing paradigms in drug discovery. In et al., M. H., editor, *Proc. Int. Beilstein Workshop*, pages 51 – 72, Berlin. Logos-Verlag.

Kuhn, H. (1955). The hungarian method for the assignment problem. *Naval Res. Logist. Quart.*, 2:83 – 97.

Kuhn, H. and Tucker, A. (1951). Nonlinear programming. In *Proc. 2nd Berkely Symposium on Mathematical Statistics and Probabilistics*, pages 481 – 492, Berkley. University of California Press.

Kwok, J. (2000). The evidence framework applied to support vector machines. *IEEE Transactions on Neural Networks*, 11(5):1162 – 1173.

Lai, P. and Fyfe, C. (2000). Kernel and nonlinear canonical correlation analysis. *Int. Journal of Neural Systems*, 10(5):365 – 377.

Lanckriet, G., Cristianini, N., Bartlett, P., Ghaoui, L. E., and Jordan, M. (2004). Learning the kernel matrix with semidefinite programming. *J. Machine Learning Research*, 5:27 – 72.

Law, M. and Kwok, J. (2001). Bayesian support vector regression. In *Proc. 11th Int. Workshop on AI and Statistics (AISTATS 2001)*, pages 239 – 244.

Lee, S., Hur, J., and Kim, Y. (2004). A graph-theoretic modeling on go space for biological interpretation on gene clusters. *Bioinformatics*, 20(3):381–388.

Leslie, C., Kuang, R., and Eskin, E. (2004). Inexact matching string kernels for protein classification. In Schölkopf, B., Tsuda, K., and Vert, J.-P., editors, *Kernel Methods in Computational Biology*, pages 95 – 112. MIT Press, Cambridge, MA.

Löfberg, J. (2004). YALMIP : A toolbox for modeling and optimization in MATLAB. In *Proceedings of the CACSD Conference*, Taipei, Taiwan. Available from http://control.ee.ethz.ch/~joloef/yalmip.php.

Lord, P., Stevens, R., Brass, A., and Goble, C. (2002). Semantic similarity measures across the gene ontology: the relationship between sequence and annotation. *Bioinformatics*, 19:1275–1283.

Luxburg, U., Bousquet, O., and Schölkopf, B. (2004). A compression approach to support vector model selection. *J. Machine Learning Research*, 5:293 – 323.

MacKay, D. (1992). The Evidence Framework Applied to Classification Networks. *Neural Computation*, 4(5):720 – 736.

MacKay, D. (1997). Gaussian Processes - A Replacement for Supervised Neural Networks? In *Proc. Neural Inf. Proc. Syst.* Lecture note.

Maggiora, G. and Shanmugasundaram, V. (2004). Molecular similarity measures. In Bajorath, J., editor, *Chemoinformatics*, pages 1 – 50. Humana Press.

Mandagere, A. and Jones, B. (2003). Prediction of bioavailability. In van de Waterbeemd, H., Lennernäs, H., and Artursson, P., editors, *Drug Bioavailability*, pages 444 – 460. Wiley-VCH, Weinheim.

Martin, Y. (1998). Pharmacophore mapping. In Martin, Y. and Willett, P., editors, *Designing Bioactive Molecules*, pages 121–148. Oxford University Press.

McCulloch, W. and Pitts, W. (1943). A logical calculus of ideas immanent in nervous activity. *Bulletin of Mathematical Biophysics*, 5:115 – 133.

Mehlhorn, K. and Näher, S. (1999). *The LEDA Platform of Combinatorial and Geometric Computing*. Cambridge University Press.

Meila, M. and Shi, J. (2001a). Learning segmentation by random walks. In *Adv. Neural Inf. Proc. Syst. 13*, pages 873–879.

Meila, M. and Shi, J. (2001b). A random walks view of spectral segmentation. *AI and Statistics (AISTATS)*.

Mercer, J. (1909). Functions of positive and negative type and their connection with the theory of integral equations. *Philosophical Transactions of the Royal Society*, A 209:415 – 446.

Minsky, M. and Papert, S. A. (1969). *Perceptrons: An Introduction into Computational Geometry*. MIT Press, Cambridge, MA.

Momma, M. and Bennett, K. (2002). A pattern search method for model selection of support vector regression. In *SIAM Conf. on Data Mining*.

Naundorf, B., Geisel, T., and Wolf, F. (2004). Dynamical response properties of a canonical model for type-i membranes. *Neural Comp.* accepted.

Ng, A., Jordan, M., and Weiss, Y. (2002). On spectral clustering: Analysis and an algorithm. In *Adv. Neural Inf. Proc. Syst. 14*.

Norinder, U. and Haeberlein, M. (2003). Calculated molecular properties and multivariate statistical analysis in absorption prediction. In van de Waterbeemd, H., Lennernäs, H., and Artursson, P., editors, *Drug Bioavailability*, pages 358 – 405. Wiley-VCH, Weinheim.

Oprea, T. I., Zamora, I., and Ungell, A.-L. (2002). Pharmacokinetically based mapping device for chemical space navigation. *J. Comb. Chem.*, 4:258–266.

Palm, K., Stenburg, P., Luthman, K., and Artursson, P. (1997). Polar molecular surface properties predict the intestinal absorption of drugs in humans. *Pharm. Res.*, 14:586 – 571.

Raedt, L. D. and Kramer, S. (2001). Feature construction with version spaces for biochemical application. In *Proc. 18th Int. Conf. on Machine Learning*, pages 258 – 265.

Rakotomamonjy, A. (2003). Variable selection using svm based criteria. *J. Machine Learning Research: Special Issue on Variable and Feature Selection*, 3:1357 – 1370.

Rarey, M. and Dixon, S. (1998). Feature trees: A new molecular similarity measure based on tree-matching. *J. Computer-Aided Molecular Design*, 12:471 – 490.

Raymond, J., Gardiner, E., Willett, P., and Rascal, P. (2002). Calculation of graph similarity using maximum common edge subgraphs. *The Computer Journal*, 45(6):631 – 644.

Resnik, P. (1995). Using information content to evaluate semantic similarity in a taxonomy. In *Proceedings of the 14th International Joint Conference on Artificial Intelligence*, volume 1, pages 448–453, Montreal.

Robinson, P., Wollstein, A., U., B., and B., B. (2003). Ontologizing gene-expression microarray data: characterizing clusters with gene ontology. *Bioinformatics*, 20(6):979–981.

Rosenblatt, F. (1958). The Perceptron: a Probalistic Model for Information Storage and Organization in the Brain. *Psychol. Review*, 65:386 – 408.

Rousseeuw, P. (1987). Silhouettes: a graphical aid to the interpretation and validation of cluster analysis. *J. Comp. and Applied Mathematics*, 20:53–65.

Russel, S. and Norvig, P. (1995). *Artificial Intelligence - A Modern Approach.* Prentice Hall Inc., New Jersey.

Salcedo-Sanz, S., Prado-Cumplido, M., Perez-Cruz, F., and Bousono-Calzon, C. (2002). Feature Selection via Genetic Optimization. In *Proc. Int. Conf. Artifical Neural Networks 2002*, pages 547 – 552.

Schölkopf, B., Burges, C., and Vapnik, V. (1995). Extracting support data for a given task. In Fayyad, U. N. and Uthurusamy, R., editors, *First Int. Conf. for Knowledge Discovery and Data Mining*, Menlo Park. AAAI Press.

Schölkopf, B., Smola, A., Williamson, R., and Bartlett, P. (2000). New support vector algorithms. *Neural Computation*, 12:1207 – 1245.

Schölkopf, B. and Smola, A. J. (2002). *Learning with Kernels.* MIT Press, Cambridge, MA.

Schölkopf, B., Tsuda, K., and Vert, J.-P. (2004). *Kernel Methods in Computational Biology.* MIT Press, Cambridge, MA.

Shah, N. and Fedoroff, N. (2004). CLENCH: a program for calculating Cluster ENriCHment using Gene Ontology. *Bioinformatics*, 20(7):1196–1197.

Shawe-Taylor, J. and Cristianini, N. (2004). *Kernel Methods for Pattern Analysis*. Cambridge University Press, Cambridge, UK.

Shi, J. and Malik, J. (2000). Normalized cuts and image segmentation. *IEEE Transactions on Pattern Analysis and Machine Intelligence*, 22(8):888–905.

Smola, A. and Schölkopf, B. (1998). A tutorial on support vector regression. Technical Report NC2-TR-1998-030, NeuroCOLT2 Technical Report Series.

Speer, N., Fröhlich, H., Spieth, C., and Zell, A. (2005a). Functional grouping of genes using spectral clustering and gene ontology. In *Proc. Int. Joint Conf. Neural Networks*, pages 298 – 303.

Speer, N., Fröhlich, H., Spieth, C., and Zell, A. (2005b). Functional distances for genes based on go feature maps and their application to clustering. In *Proc. IEEE Symp. on Comp. Intel. in Bioinf. and Comp. Biology (CIBCB 2005)*, pages 142 – 149, San Diego, USA. IEEE Press.

Speer, N., Spieth, C., and Zell, A. (2004). A memetic clustering algorithm for the functional partition of genes based on the Gene Ontology. In *Proceedings of the IEEE Symposium on Computational Intelligence in Bioinformatics and Computational Biology*, pages 252–259.

Speer, N., Spieth, C., and Zell, A. (2005c). Spectral clustering gene ontology terms to group genes by function. In *Proceedings of the 5th Workshop on Algorithms in Bioinformatics (WABI 2005)*, volume 3692 of *Lecture Notes in Bioinformatics (LNBI)*, pages 001–012. Springer.

Sturm, J. (1999). Using sedumi1.02 a matlab toolbox for optimization over symmetric cones. *Optimization Methods and Software*, 11/12(1 - 4):625 – 653.

The Gene Ontology Consortium (2004). The gene ontology (GO) database and informatics resource. *Nucleic Acids Research*, 32:D258–D261.

Tikhonov, A. and Arsenin, V. (1977). *Solutions of ill-posed problems*. W.H. Winston.

Todeschini, R. and Consonni, V., editors (2000). *Handbook of Molecular Descriptors*. Wiley–VCH, Weinheim.

Vafaie, H. and Jong, K. D. (1998). Evolutionary feature space transformation. In Liu, H. and Motoda, H., editors, *Feature Extraction, Construction and Selection: a data mining perspective*, pages 307 – 323. Kluwer.

van de Waterbeemd, H. and Gifford, E. (2003). ADMET *In Silico* Modelling: Towards Prediction Paradise? *Nature Reviews: Drug Discovery*, 2:192–204.

Vandenberghe, L. and Boyd, S. (1996). Semidefinite programming. *SIAM Review*, 38(1):49 – 95.

Vapnik, V. (1979). *Estimation of Dependencies Based on Empirical Data (in Russian)*. Nauka, Moscow.

Vapnik, V. (1995). *The Nature of Statistical Learning Theory*. Springer, New York, NY.

Vapnik, V. (1998). *Statistical Learning Theory*. John Wiley and Sons, New York.

Vapnik, V. and Chapelle, O. (2000). Bounds on error expectation for svm. In *Advances in Large Margin Classifiers*, pages 261 – 280. MIT Press, Cambridge, MA.

Vapnik, V. and Chervonenkis, A. (1968). Uniform convergence of frequencies of occurences of events to their probabilities. *Dokl. Akad. Nauk SSSR*, 181:915 – 918.

Verma, D. and Meila, M. (2003). A comparison of spectral clustering algorithms. Technical Report 03-05-01, University of Washington, CSE.

Vishwanathan, S. and Smola, A. (2004). Fast Kernels for String and Tree Matching. In Schölkopf, B., Tsuda, K., and Vert, J.-P., editors, *Kernel Methods in Computational Biology*, pages 113 – 130. MIT Press, Cambridge, MA.

Volgushev, M. and Eysel, U. (2000). Noise makes sense in neuronal computing. *Science*, 290.

Volgushev, M., Pernberg, J., and Eysel, U. (2003). Gamma-frequency fluctuations of the membrane potential and response selectivity in visual cortical neurons. *Eur. J. Neurosci.*, 17:1768 – 1776.

Washio, T. and Motoda, H. (2003). State of the art of graph-based data mining. *SIGKDD Explorations Special Issue on Multi-Relational Data Mining*, 5.

Wegner, J., Fröhlich, H., and Zell, A. (2003a). Feature selection for Descriptor based Classification Models: Part II - Human Intestinal Absorption (HIA). *J. Chem. Inf. Comput. Sci.*, 44:931 – 939.

Wegner, J., Fröhlich, H., and Zell, A. (2003b). Feature Selection for Descriptor based Classificiation Models: Part I - Theory and GA-SEC Algorithm. *J. Chem. Inf. Comput. Sci.*, 44:921 – 930.

Wegner, J. K. (2006). *Data Mining und Graph Mining auf molekularen Graphen - Cheminformatik und molekulare Kodierungen für ADME/Tox & QSAR-Analysen*. PhD thesis, Eberhard-Karls Universität Tübingen.

Weiss, Y. (1999). Segmentation using eigenvectors: A unifying view. In *ICCV (2)*, pages 975–982.

Wessel, M. D., Jurs, P. C., Tolan, J. W., and Muskal, S. M. (1998). Prediction of Human Intestinal Absorption of Drug Compounds from Molecular Structure. *J. Chem. Inf. Comput. Sci.*, 38:726 – 735.

Weston, J. (1999). *Extensions to the Support Vector Method*. PhD thesis, Royal Holloway University of London.

Weston, J., Elisseeff, A., Schölkopf, B., and Tipping, M. (2002). Use of the zero-norm with linear models and kernel methods. *J. Machine Learning Research Special Issue on Variable and Feature Selection*, 3:1439 – 1461.

Weston, J., Mukherjee, S., Chapelle, O., Pontil, M., Poggio, T., and Vapnik, V. (2001). Feature selection for SVMs. In Solla, S., Leen, T., and Müller, K.-R., editors, *Adv. Neural Inf. Proc. Syst. 13*. MIT Press.

Weston, J. and Watkins, C. (1999). Multi-class support vector machines. In Verleysen, M., editor, *Proc. Europ. Symp. Artificial Neural Networks*, Brussles.

Willams, C. (1997). Prediction with gaussian processes: From linear regression to linear prediction and beyond. Technical Report NRG/97/012, Aston University, UK.

Wolfe, P. (1961). A duality theorem for nonlinear programming. *Quarterly of Applied Mathematics*, 19:239 – 244.

Xue, Y., Li, Z. R., Yap, C. W., Sun, L. Z., Chen, X., , and Chen, Y. Z. (2004). Effect of molecular descriptor feature selection in support vector machine classification of pharmacokinetic and toxicological properties of chemical agents. *J. Chem. Inf. Comp. Sci.*, 44(5):1630 – 1638.

y Arcas, B. A., Fairhall, A., and Bialek, W. (2003). Computation in a single neuron: Hodgkin and huxley revisited. *Neural Comp.*, 15:1715 – 1749.

Yamanishi, Y., Vert, J.-P., and Kaneshisa, M. (2004a). Heterogenous data comparison and gene selection with kernel canonical correlation analysis. In Schölkopf, B., Tsuda, K., and Vert, J.-P., editors, *Kernel Methods in Computational Biology*, pages 209 – 230. MIT Press, Cambridge, MA.

Yamanishi, Y., Vert, J.-P., and Kaneshisa, M. (2004b). Protein network inference from multiple genomic data: a supervised approach. *Bioinformatics*, 20(1):i363 – i370.

Yazdanian, M., Glynn, S., Wright, J., and Hawi, A. (1998). Correlating partitioning and caco-2 cell permeability of structurally diverse small molecular weight compounds. *Pharm. Res.*, 15:1490 – 1494.

Yee, S. (1997). In vitro permeability across caco-2 cells (colonic) can predict in vivo (small intestinal) absorption in man - fact or myth. *Pharm. Res.*, 14:763 – 766.

Yoshida, F. and Topliss, J. (2000). QSAR model for drug human oral bioavailability. *J. Med. Chem.*, 43:2575 – 2585.

Yu, L. and Liu, H. (2004). Efficient feature selection via analysis of relvance and redundancy. *J. Machine Learning Research*, 5:1205 – 1224.

Zeeberg, B., Feng, W., Wang, G., and *et al.*, A. F. (2003). GOminer: a resource for biological interpretation of genomic and proteomic data. *Genome Biology*, 4(R28).